NATION AND NURTURE IN SEVENTEENTH-CENTURY ENGLISH LITERATURE

Nation and Nurture in Seventeenth-Century English Literature

RACHEL TRUBOWITZ

OXFORD

UNIVERSITY PRESS

OXFORD
UNIVERSITY PRESS

Great Clarendon Street, Oxford, OX2 6DP,
United Kingdom

Oxford University Press is a department of the University of Oxford.
It furthers the University's objective of excellence in research, scholarship,
and education by publishing worldwide. Oxford is a registered trade mark of
Oxford University Press in the UK and in certain other countries

Published in the United States of America by Oxford University Press
198 Madison Avenue, New York, NY 10016, United States of America

British Library Cataloguing in Publication Data
Data available

Library of Congress Control Number: 2012002800

ISBN 978-0-19-960473-9

For my mother and in memory of my father

Acknowledgments

Many people have supported the research and writing of this book, and it is my pleasure to thank them.

I am grateful to Andrew McNeillie, who was the first to endorse and encourage my project at Oxford University Press, and to Jacqueline Baker, who took over the book after he left. Her patience and professionalism have been indispensable at every stage of the process. Many thanks as well are owed to Ariane Petit for her unstinting helpfulness, even with minute details. Sarah Cheeseman's excellent copyediting saved me from many errors. I also am indebted to the two anonymous readers for the press, from whose expert suggestions I have benefitted enormously.

I thank Burt Feintuch and the Center for the Humanities at the University of New Hampshire for a senior research fellowship, which, in combination with a sabbatical leave from the university, allowed me to spend a year conducting research at the beginning of this project.

Among many friends and colleagues, I wish especially to acknowledge Achsah Guibbory and Paul Stevens for the extraordinary quality of their support. Richard Strier and Anne Lake Prescott gave me early encouragement. Mihoko Suzuki's faith in my project was sustaining. Tom Luxon consistently cheered me on. Jason Rosenblatt helped me at a crucial juncture. For their appreciation of my work over the years, I thank John Michael Archer, Ann Baynes Coiro, Dayton Haskins, Marshall Grossman (in memory), Laura Lunger Knoppers, Barbara K. Lewalski, John Rogers, Elizabeth Sauer, and Edward W. Tayler. I am grateful to Sharon Achinstein, Jim Holstun, Catherine Gimelli Martin, Naomi Miller, Naomi Yavneh, Mihoko Suzuki, and Christina Malcolmson for including my work (preliminary to this book) in their edited volumes. Jackie DiSalvo, Achsah Guibbory, Paul Stevens, Linda Gregerson, Laura Knoppers, Diana Henderson, Marina Leslie, David Loewenstein, Elizabeth Sauer, and Joseph Wittreich provided pivotal opportunities for me to present my work at MLA, RSA, the Seminar on Women and Gender in Reformation Europe at Harvard, the Canada Milton Seminar, the Northeast Milton Seminar, and the International Milton Symposium. I am grateful to the members of the Northeast Milton Seminar for their invaluable feedback on an early draft of the *Samson* chapter.

My UNH colleagues (past and present) have been incredibly supportive, especially Elizabeth Hageman, Douglas Lanier, Sarah Sherman, Diane Freedman, Paula Salvio, Brigitte Bailey, Michael Ferber, and the late

Michael V. Deporte. I have been lucky to work collaboratively with Elizabeth Jane Bellamy. Special thanks are owed to Mary Beth Rhiel for her wisdom and generosity and to Sandhya Shetty, who read every word of my manuscript (more than once).

I wish to acknowledge my cherished friends: Eriko Amino, Kate Ballen, Jimmy Roberts, Simon Ross, and Rita Costabile Tobin.

Above all, I thank my family for their love and support: my brother, Eugene, my amazing children, Ella and Simon, my mother, Ethel, and my father, Sidney (in memory). My book is dedicated to my parents.

Some of my chapters draw on materials published elsewhere in different forms. I am grateful to the various presses for granting me permission to reprint them in this book.

A section of Chapter 2 appeared as "Cross-Dressed Women and Natural Mothers: 'Boundary Panic' in *Hic Mulier*," in *Debating Gender in Early Modern England, 1500–1700*, eds. Christina Malcolmson and Mihoko Suzuki (New York: Palgrave Macmillan, 2002), 185–208, reprinted by permission of Palgrave Macmillan.

A section of Chapter 4 includes materials published in "Female Preachers and Male Wives: Gender and Authority in Civil War England," *Prose Studies*, special ed. James Holstun, 14.2 (1991), 112–33, reprinted by permission of the Taylor and Francis Group.

An early version of Chapter 5 was published as " 'I was his nursling once': Nation, Lactation and the Hebraic in *Samson Agonistes*," in *Milton and Gender*, ed. Catherine Gimelli Martin (Cambridge: Cambridge University Press, 2004), 167–83, reprinted by permission of Cambridge University Press.

Table of Contents

List of Images

*This image is to be used for the book cover.

List of Abbreviations

CPW   John Milton, *Complete Prose Works*, gen.ed. Don M. Wolfe, 8 vols. (New Haven, CT: Yale University Press, 1953–82). This edition is used for all Milton's prose.

CPEP   John Milton, *The Complete Poetry and Essential Prose of John Milton*, eds. William Kerrigan, John Rumrich, and Stephen M. Fallon (Modern Library-Random House, 2007). This edition is used for all Milton's poetry.

EEBO   *Early English Books Online* (Chadwick & Healey, 2011).

ODNB   *Oxford Dictionary of National Biography* (Oxford: Oxford University Press, 2004–9; online versison).

Introduction

In William Hoyle's engraved frontispiece to Michael Drayton's *Poly-Olbion,*[1] Great Britain is allegorized as a goddess-mother, dressed in a map and holding a scepter in her right hand (see figure 1).

With her left arm, she serenely cradles a cornucopia of fruits, which, like a baby, appears to be nursing at her breast. Around the edges of the frontispiece and flanking the maternal goddess are small statue-like figures of Britain's forefathers: Brutus, Caesar, the Saxon Hengist, and William the Conqueror. While these four male monarchs look out warily and competitively from the margins, the nursing goddess calmly occupies the center. They are marble, inert, and, ultimately, superfluous. She is natural, fertile, reproductive, and succoring, capable of growth and expansion, yet self-sufficient and unified. Drayton dedicates his poem to the future King Henry, who died in 1612, the same year that the first installment of *Poly-Olbion* with Hoyle's frontispiece was published. The frontispiece, how-ever, belies the dynastic promise of the dedication. The rival kings represent Britain's past incarnations (Trojan, Roman, Saxon, and Norman) as anxious, discontinuous, sterile, and outmoded. In contrast, the central maternal figure is whole, solid, and full of national promise and natural plenitude. As Richard Helgerson points out, "So satisfying is this image, so whole and so right, that it makes any suggestion of an underlying ideo-logical struggle seem forced . . . Drayton's frontispiece presents the results of a conceptual revolution as though nothing had happened at all."[2]

Although Helgerson does not identify the maternal goddess as a nurs-ing or nurturing mother, it seems impossible to ignore this surprisingly under-examined dimension of the frontispiece. The focal point is the ma-ternal goddess's bared left breast, at which a cornucopia of fruits suckles.[3]

---

[1] Michael Drayton, *Poly-Olbion* (London, 1612), frontispiece.

[2] Richard Helgerson, *Forms of Nationhood: The Elizabethan Writing of England* (Chicago: University of Chicago Press, 1992), p. 122.

[3] The exposure of one breast has Amazonian implications. See Katherine Schwarz, "Missing the Breast: Desire, Disease, and the Singular Effect of Amazons," in *Body in Parts: Fantasies of Corporeality in Early Modern Europe*, ed. David Hillman and Carla Mazzio (New York: Routledge, 1997), pp. 147–69.

Fig. 1. Michael Drayton, frontispiece, *Poly-Olbion*, 1612.

Readers are invited to identify with the fruits in the cornucopia: as lapping up social nourishment from the breast of the mother country. Through Drayton's serene mother-goddess, the intimate relationship between nursing mother and child effortlessly expands to fit the national whole. As a loving, nursing mother, Great Britain contentedly masks the contentiousness of this historical moment, as conflicts between court and country, anti- and pro-Unionists, among others, rapidly intensify. By rendering conspicuous political and religious conflicts invisible, Drayton's nursing Great Britain emblematically plays a crucial consolidating national role. The nursing mother (and nursing father) as national mediator is a recurrent trope in many of the texts to be considered in this study.

As Drayton's frontispiece suggests, the nursing mother gains cultural capital as a satisfying symbol of national wholeness—capital that a wide range of rival political parties and religious sects wish to acquire and spend. As I shall demonstrate in my first chapter, Spenser's *A View on the Present State of Ireland* is one of the most important early modern texts to associate mother's milk with the mother tongue: maternal nursing helps to naturalize the emerging language-based nation. For Spenser, the enduring practice among the Anglo-Irish gentry and nobility of hiring Irish wet nurses hinders their children's acquisition of the English language and English identity as they mature. Because of their primal attachments to their wet nurses, Anglo-Irish children remain forever divided in their love and loyalties between Ireland and England. English mothers in Ireland thus should nurse their own babies, rather than give them over to Irish wet nurses. By drinking in the English language with their own mother's milk, English children in Ireland will lap up their mother tongue and love their mother country from the breast. No longer will their loyalties be divided between England and its Irish colony.[4]

In *Richard II*, John of Gaunt, in an oft-cited speech, also depicts England as a loving nurse: "this England/This nurse" (2.1. 50–1).[5] Rather than a linguistic fount of Englishness, however, Gaunt's nursing England is a bountiful garden-kingdom. But, as his elegiac speech also makes clear, his luscious Edenic images allude to a national era that has passed. England's nurturing "demi-paradise" (2.1. 42) is a paradise lost. What is lost, more specifically, is the nation's rootedness both in the land and in the blood and body of its hereditary monarch. Land and ruler split apart from one another, generating civil strife, rebellion, and regicide. Rather than nourish its "happy breed of men" (2.1. 45) and its heroic kings, "Feared by their breed" (2.1. 52), England gorges on and is choked by its own overflowing bounty: "With eager feeding good doth choke the feeder" (2.1. 37). As Gaunt prophesies: "That England that was wont to conquer others/Hath made a shameful conquest of itself" (2.1. 65–6). The gap that opens up in *Richard II* between the dynastic kingdom and the land-based nation is the prelude to civil war. John of Gaunt's association of England's lost *integritas* with the nursing mother poignantly evokes the past, but it fails to address the present troubles, and so, despite his eloquence, his pacific desire to restore the nation-as-nurse to its former wholesomeness and health goes unattended.

---

[4] Edmund Spenser, *A View of the Present State of Ireland*, ed. W.L. Renwick (Oxford: Clarendon, 1970).

[5] Quotations from Shakespeare's plays are taken from *The Norton Shakespeare*, eds. Stephen Greenblatt, Walter Cohen, Jean E. Howard, Katharine Eisaman Maus (New York: W.W. Norton, 1997). They are noted in the text.

What is metaphor or allegory in *Richard II* is scientific fact in Thomas Phaer's *The Book of Children*, the first published text in English on pediatrics. Phaer considers the nurse and her milk to be indispensable to the physical health of the child, the wholesomeness of his character as an adult, and the larger law and order of the kingdom.[6] Citing Virgil, Phaer attributes tyranny to bad nurses:

> I entend to write somewhat of the nurse and of the milke, with qualities and complexions of the same, for in it consisteth the chief point or sum not onely of the maintenance of health, but also of the fourming or infecting either of the wit or maners, as the Poet Virgil, when he would describe an uncourteous, churlishe, and rude condishioned Tyrant, did attribute the faulte unto the giver of the milke.[7]

For Phaer, as for Drayton and John of Gaunt, the nurse or nursing mother is essential to the nation's health and well-being.

I

As Drayton's frontispiece, Phaer's pediatrics text, and Shakespeare's *Richard II* exemplify, early modern conceptions of nursing and maternal nurture find their way into a variety of different political conversations, especially those concerning national identity. Although the revaluation of motherhood and the conceptual reformation of the English nation have been studied extensively, they mostly have been investigated in isolation from one another. Surprisingly, little has yet been said about the mixing and mingling of rhetoric about the nation and nursing—an omission that *Nation and Nurture in Seventeenth-Century English Literature* attempts to redress. As I shall demonstrate, the conceptual reformation of the nation and the revaluation of maternal nursing take place simultaneously.[8] Maternal nurture newly occupies a central if highly contested place in the early modern cultural imagination at the precise moment when England undergoes a major conceptual paradigm shift: from the old dynastic body politic, organized by organic bonds of blood, soil, and kinship, to the

---

[6] In all likelihood, Phaer was responsible for the introduction of a bill "for the nursing of Children in Wales" in the 1547 parliament. Philip Schwyzer, "Phaer, Thomas (1510?–1560)," *Oxford Dictionary of National Biography*; online ed. October 2009 [http://www.Oxforddnb.com/view/article/22085].

[7] Thomas Phayer [Phaer], *The kegiment[sic] of life whereunto is added a treatyse of the pestilence, with The book of children newly corrected and enlarged by T. Phayer* (London: Edward Whitchurche 1594) sig. Kr. STC 2nd ed./11969. *EEBO*. In 1544 Phaer first published his *Boke of Chyldren* and two other medical works of his own in one volume with his translation of Jean Goeurot's *The Regiment of Life*.

[8] Helgerson's magnificent *Forms of Nationhood* does not treat gender.

new, post-dynastic, modern nation, comprised of disembodied, symbolic, and affective relations.[9] Whereas the old organic nation categorically differentiates its native population by biological pedigree (bloodline and birthright), the new early modern nation reinforces its abstract unity as a political entity through what Claire McEachern terms "a rhetoric of social intimacy."[10] Over the last three decades, *pace* Benedict Anderson, many magnificent studies have investigated this shift between the organic paradigm and the abstract or "imagined" one.[11] No attention, however, has been paid as yet to the crucial role that the figure of the reformed nursing mother plays in translating the old nation into the new one. As I shall document, the emphasis on intimacy and affective unity in the political rhetoric of the nation coincides historically with a new emphasis on maternal nurture and nursing in the domestic rhetoric of the private household. The hitherto unexamined convergences between these two discourses—and the changeable mix of cultural and political ambitions that they reflect—find vivid expression in the texts analyzed in my study, both traditional writings, such as James I's *Basilikon Doron* and reformed ones, such as John Milton's *Samson Agonistes*.

My investigation of maternal nurture and the nation addresses two additional issues. First, I aim to re-examine early modern England as a

---

[9] In *The Year 2000* (New York: Pantheon, 1983), Williams emphasizes the need to distinguish between the ancient nation, which is ethnic or tribal, and the modern nation-state, which is entirely artificial. See Christopher Hill, "The Protestant Nation," in his *Collected Essays II: Religion and Politics in Seventeenth-Century England* (Brighton, 1986) and "National Unification" in his *Reformation to Industrial Revolution: The Making of Modern English Society, 1530–1780* (New York: Pantheon, 1968); Tom Nairn, *The Break-up of Britain: Crisis and Neo-Nationalism* (London: Verso, 1981); Benedict Anderson, *Imagined Communities: Reflections of the Origin and Spread of Nationalism* (London: Verso, 1983); Claire McEachern, *The Poetics of English Nationhood, 1590–1612* (Cambridge: Cambridge University Press, 1996); and *Early Modern Nationalism and Milton's England*, eds. David Loewenstein and Paul Stevens (Toronto: University of Toronto Press, 2008).

[10] McEachern, p. 16. Important studies of early modern literature and nationalism include Paul Stevens, "Milton's Janus-Faced Nationalism: Soliloquy, Subject, and the Modern Nation State," *Journal of German and English Philology*, 100 (2001): 247–68; Claire McEachern, *The Poetics of Nationhood*; Linda Gregerson, "Colonials Write the Nation: Spenser, Milton, and England on the Margins," in *Milton and the Imperial Vision*, eds. Balachandra Rajan and Elizabeth Sauer (Pittsburgh: Duquesne University, 1999); Richard Helgerson, *Forms of Nationhood: The Elizabethan Writing of England* (Chicago: University of Chicago Press, 1992).

[11] For arguments that challenge the identification of sixteenth- and seventeenth-century England as a modern nation, see J.G.A. Pocock, "British History: A Plea for a New Subject," *Journal of Modern History*, 7 (1975): 601–21, Linda Colley, *Britons: Forging the Nation, 1707–1837* (New Haven: Yale University Press, 1992), Collin Kidd, *British Identities Before Nationalism: Ethnicities and Nationhood in the Atlantic World, 1600–1800* (Cambridge: Cambridge University Press, 1999), and Joad Raymond, "Complications of Interest: Milton, Scotland, Ireland, and National Identity in 1649," *Review of English Studies* (2004): 315–45.

political entity from the vantage point of gender studies in its current, post-second-wave guise. In this, my book builds on the pioneering work of feminist scholars, such as Jean Howard, Phyllis Rackin, and Jody Mickalachki, who have laid the groundwork for understanding the early modern English nation in gendered terms.[12] Many of these landmark studies are informed by the theory of separate spheres. This influential theory, which dates to the 1980s, posits that, during the late sixteenth and early seventeenth centuries, the hitherto unified public-and-private realms split into two separate, but unequally valued, gendered domains: the marginal, female world of domesticity and the central male world of commerce and politics.[13] Since the early 1990s, second-wave feminist scholars have complicated the separate-spheres narrative by reviewing its binary, gendered divisions through a variety of larger cultural lenses: emerging capitalism, early acts of colonization, the shift from a Mediterranean to an Atlantic economy, household labor, commerce and consumerism, Puritan emphasis on interiority, and England's increasing contact with foreign cultures, among others.[14] From this second-wave vantage point, the separateness between the male public realm and female private domain posited by the theory of separate spheres does not stand up to new feminist-historical documentation of material practices in the seventeenth century.[15]

---

[12] Jean E. Howard and Phyllis Rackin, *Engendering A Nation: A Feminist Account of Shakespeare's English Histories* (London and New York: Routledge, 1997); Jodi Mikalachki, *The Legacy of Boadicea: Gender and Nation in Early Modern England* (London: Routledge, 1998); Megan Matchinski, *Writing, Gender, and State in Early Modern England: Identity Formation and the Female Subject* (Cambridge: Cambridge University Press, 1998); Mihoko Suzuki, *Subordinate Subjects: Gender, the Political Nation, and Literary Form in England, 1588–1688* (Burlington, VT: Ashgate, 2002); and Wendy Wall, *Staging Domesticity: Household Work and English Identity in Early Modern Drama* (Cambridge, Cambridge University Press, 2003).

[13] Joan Kelly's landmark essay, "Did Women Have a Renaissance?" in *Women, History, and Theory: The Essays of Joan Kelly* (Chicago: University of Chicago Press, 1984), pp. 19–50, offers the most influential formulation of this idea. See also Alice Clark's *Working Life of Women in the Seventeenth Century* (1919; reprint, London: Routledge and Kegan Paul, 1982). Recent applications of the separate-spheres theory can be found in Thomas A. Laqueur, *Making Sex: Body and Gender from the Greeks to Freud* (Cambridge: Cambridge University Press, 1990); Elizabeth Kowaleski-Wallace, *Consuming Subjects: Women, Shopping, and Business in the Eighteenth Century* (New York: Columbia University Press, 1997); and Jodi Mikalachki, *The Legacy of Boadicea: Gender and Nation in Early Modern England* (London: Routledge, 1998).

[14] Important challenges have been made to the separate-spheres theory by Amy Louise Erickson, *Women and Property in Early Modern England* (London: Routledge, 1993), and Robert B. Shoemaker, *Gender in English Society, 1650–1850: The Emergence of Separate Spheres* (London: Addison Wesley Longman Limited, 1998).

[15] See Wendy Wall, *Staging Domesticity: Household Work and English Identity in Early Modern Drama* (Cambridge: Cambridge University Press, 2002); Natasha Korda, *Shakespeare's Domestic Economies: Gender and Property in Early Modern England* (Philadelphia: University of Pennsylvania Press, 2002); and Lena Cowen Orlin, *Private Matters and Public Culture in Post-Reformation England* (Ithaca: Cornell University Press, 1994).

Despite its inroads into the separate-spheres theory, however, the 1990s fluid, private–public view of gender has been unable to dislodge completely the still-influential 1980s paradigm of fixed early modern public–private, male–female boundaries.[16] This may be because, despite their overt differences, the two views in fact complement each other. Although there is little material evidence to support the separate-spheres theory, the theory nevertheless accurately accounts for seventeenth-century cultural perceptions of the relations (which differ from the facts) between the two genders. As Mary Beth Rose persuasively argues, gender is a fluid-and-fixed, integrated-and-separate, real-and-imaginary, private-public construct in the early modern period. I take Rose's argument that male and female, public and private, are at once distinct and overlapping early modern cultural categories as my starting point for what I contend is the mostly overlooked convergence between changing perceptions of motherhood, on the one hand, and shifting paradigms of the nation, on the other hand. My first aim is to demonstrate that women are written out of public life and consigned to the devalued female sphere of the private household, *and* they, as newly privatized mothers, also gain considerable cultural capital as shapers and protectors of the English national character—all at the same time.[17]

The second issue is intimately related to my first. Reviewing the reformation of the nation in light of the reformation of motherhood highlights in a completely new way the Judeo-Christian implications of the old nation–new nation paradigm shift in post-Reformation England. As we shall see, these important but hitherto completely obscure implications are especially clear in Milton's prose tracts and mature poetry. I focus on *Of Education*, *Areopagitica*, *Paradise Lost* and *Samson Agonistes*. Although Milton's early shining ideal of the reformed England as the new Israel loses some of its luster between the prose of his middle years and the poetry of his maturity, in all of these texts, the poet maps the translation

---

[16] As Mary Beth Rose argues, the theory retains its powerful influence on feminist scholars of early modern England; see Introduction, *Gender and Heroism in Early Modern England* (Chicago: University of Chicago Press, 2002). Rose affirms that the theory of separate spheres fails to account for the cultural shift in early modern England from a traditional, male paradigm of action-driven heroism to a new, female paradigm of heroic suffering. But, she also emphasizes that "the reasons for changing representations of heroism are clarified when viewed as part of, rather than as running counter to, the larger cultural effort to distinguish and distinctly gender private and public spheres," p. xii.

[17] Important studies of motherhood and caregiving in early modern England include: Janet Adelman, *Suffocating Mothers: Fantasies of Maternal Origin in Shakespeare's Plays, Hamlet to The Tempest* (New York and London: Routledge, 1992), Theresa M. Krier, *Birth Passage: Maternity and Nostalgia, Antiquity to Shakespeare* (Ithaca: Cornell University Press, 2001), and Naomi J. Miller and Naomi Yavneh, eds., *Maternal Measures: Figuring Caregiving in the Early Modern Period* (Burlington, VT: Ashgate, 2000).

Fig. 2. Jan Wiemix, *The power of women*, late sixteenth century.

of the old into the new nation onto the transition between the Old and New Testaments. Simultaneously, in a series of radical appropriations, (abstract in his prose and typological in his verse), Milton substitutes the reformed or new mother for the Blessed Virgin as spiritual mediatrix, a role assigned to Mary, especially the nursing Madonna, by Roman Catholic doctrine. Milton underscores Mary's spiritual qualities. Whereas a Hebraic-Puritan equation between maternal nurture, physical nature, and national identity governs the construction of Englishness in the domestic guidebooks, in *Paradise Lost*, the new or reformed mother, "second Eve" (5.387), is a wholly spiritualized figure.[18] In her prelapsarian state, Eve is closely related to nature and biological nurture—and with the potential for an organic, one-world nation. This organic model of nation and nurture, however, is rendered obsolete by the Fall. Instead of an embodied community, "second Eve" (5.387) is associated with a spirit of

---

[18] All references to Milton's poetry are taken from *The Complete Poetry and Essential Prose of John Milton, ed.* William Kerrigan, John Rumrich, and Stephen M. Fallon (New York: The Modern Library, 2007). They are noted in the text.

unity, in which bloodline and natural birthright play no role. Whereas guidebook writers like William Gouge define the new mother as a natural nurse, Milton emphasizes her spiritual nurture. As we shall see, Mary in *Paradise Regained* is the poet's exemplar of the new spiritual mother.

Milton's typological view of providential history is distinctly gendered: it is associated with the spiritual mother of the new era and disassociated from the biological father of the old one. I argue that, to analyze Milton's gendered typological perspective of the shift from the old to the new Israel, we need to adopt a new, hybrid vantage point. Specifically, I propose to combine post-second wave feminist history and literary history with recent Jewish-studies reconsiderations of the relations between Judaism and Christianity in late antiquity. This is the period that inaugurates the two religions' self-defining battles for discursive control over the hyphen that separates and relates the Judeo- and the Christian. In my first chapter, I set the stage for this combined feminist/Jewish-studies approach to Milton in my analysis of the late sixteenth-century engraving, *The power of women* (Rotterdam, Boysmans-van Beuningen Museum) (see figure 2).[19]

Ilja Veldman describes *The power of women* as an illustration of 3 Esdras 4:15: "Women have borne the king and all the people that bear rule by sea and land." I argue that "*The power of women*" also is an example of pictorial biblical typology: Old Testament law (identified with Solomon and Samson) gives way to New Testament love and the milk of the gospel as figured by the nursing mother. I also argue that a similar typological translation of law into love and the old father into the new mother informs Milton's vision of national reformation in *Paradise Lost*.

Whether natural or spiritual, the new nursing mother is of special value to anti-traditionalists, since she is instrumental in shaping a new kind of Englishness, undetermined in any way by status, landownership, genealogy, or other customary, external markers of identity and organic measures for membership in the national whole. But the cultural worth of mother's milk also is ambiguous. On the one hand, it naturally imprints the abstract

---

[19] Ilja M. Veldman identifies "*The power of women*" (no. 3 of "*The four powers*" series) as an engraving by Jan Wiernix after Abrosius Francken; it is clearly influenced by Martin de Vos. The engraving is accompanied by a six-line poem in Latin: "The woman surpasses kings and wines. She has borne the king and the mighty of the land and sea. Reason flies in the face of her unbridled love, and spiritual virtue withdraws. See men weep when she weeps, rejoice when she is happy, become fearless when she drives out fear. Verily, woman subjugates the king." (231). Francois Borin, Madelyn Kahr, and Patricia Parker identify the image as a drawing, entitled "Allegory of the Power of Women," which is attributed to or influenced by Martin de Vos. The drawing bears a partly legible inscription, which suggests that it is a "picture" [*beleeden*] of a proverb by Sebastian Franck(e), the sixteenth-century German freethinker and radical Reformer. Franck published his *Spruchworter*, a collection of proverbs, in 1541.

virtues prized by those attempting to produce an anti-customary model of Englishness based on internal and intangible qualities, or, in a word, merit. On the other hand, however, the Galenic theory that mothers' milk is "white blood" strengthens traditional, genealogical, and bloodline paradigms of class and national identity. "Experience teacheth, that God converteth the mothers bloud into the milke where with the child is nurse," write Robert Cleaver and John Dod.[20] The figure of the nursing mother thus has enormous value both to those who wish to abolish customary measures of personal and national identity and to those who wish to preserve these same customary criteria for defining England as a body politic organized in a natural hierarchy descending from head to foot.

For reformers, the obscurity of the maternal-centered private home further enhances its cultural significance and political worth. For the radical sects, the private home gains new importance as an off-stage site for practicing dissent and fomenting revolution. As Achsah Guibbory observes, "the practices of nonconformist worship…had to take place secretly in forests, fields, caves, and especially private homes. The private home of the family became the primary site for dissenting worship."[21] Thus, in *Paradise Regained*, Jesus returns "Home to his mother's house private" (4. 639) after his secret triumph over Satan. To quell dissent, traditionalists tried to wrest control of the private home and make it serve conservative rather than innovative objectives. As we shall see in Chapter 3, James I in *Basilikon Doron* and Charles I in *Eikon Basilike* draw on reformed notions of the private home and the nursing mother in their self-representations of kingship.

Pressure thus is put on the traditional family/nation from both reformers and conservatives. Just as the customary contours of the body politic are threatened by the radicalized alternity of the female-centered Puritan household, so they also are threatened from the centers of official Church-and-State power by the desire for national expansion, territorial and intellectual. When she complements the status quo, then the nursing mother is associated with embodied blood-and-soil measures of social entitlement. But when she challenges customary social norms, the nursing mother is identified with the rhetoric of social intimacy and the abstract terms of affective nationalism. This very same reformed discourse of spiritual community enables the abstract modern nation to prevail over—but

---

[20] Robert Cleaver and John Dod, *A Godly Form of Household Government* (London, 1621. 1st edn, 1614), sig. 54v.

[21] Achsah Guibbory, *Ceremony and Community from Herbert to Milton: Literature, Religion, and Cultural Context in Seventeenth-Century England* (Cambridge: Cambridge University Press, 1998), p. 190.

not wipe out entirely—the traditional dynastic nation, organized by natural bonds of blood, genealogy, and kinship. A satisfying symbol of unity, on the one hand, the reformed mother also reflects the shifting political and moral fault lines of seventeenth-century English culture, on the other hand.

The battle between seventeenth-century anti-monarchists and the Stuart monarchy over national definition puts the period's conflictual paradigm of nursing motherhood into a more precise historical context.[22] As has been well documented, British imperialism in its first phase was partly inaugurated by James I's ultimately unsuccessful campaign for Union. The king's desire to federate England and Scotland was energized by his vision of the nation as a new, greater Britannia. This vision conspicuously differed from the Elizabethan ideal of England as an insular and intact kingdom. Nostalgia for Elizabeth I in combination with strong anti-Scottish sentiment thwarted James's plans for a greater Britain. These plans also encountered considerable resistance from Parliament, which sought to limit the king's prerogative. Like James, Charles I battled the MPs' attempts to restrain his royal sovereignty. In 1629 he dissolved Parliament (until 1640). His long Personal Rule and his absolutist vision of his kingdom met with strong opposition from a variety of dissenting parties: radical sectarians, republicans, and regicides, among others.[23] In his regicidal prose tracts of 1643–45, Milton depicts Charles as an unfeeling tyrant, who in placing himself above the law, breaks his covenantal relationship with his people—a capital offense.[24] Milton also contrasts the morbidity of the Stuart kingdom with the vitality of the newly reformed England. In *Areopagitica*, he envisions England as freshly arisen, young, and full of promise—sent aloft by its resurrected spirit rather than weighed down, as is the old dynastic nation, by state illnesses and corruption.[25] But, even with the breathtaking sublimity of his prose, Milton's heroic efforts to (like Samson) knock down the conceptual frameworks of the

[22] Robert Cust and Ann Hughes, eds., *Conflict in Early Stuart England: Studies in Religion and Politics, 1603–42* (London and New York: Longman, 1989), and Kevin Sharpe and Peter Lake, eds., *Culture and Politics in Early Stuart England* (Palo Alto: Stanford University Press, 1994). For revisionist perspectives, see Conrad Russell, ed., *The Origins of the English Civil War* (New York: Barnes & Noble, 1973) and *The Causes of the English Civil War* (Oxford: Clarendon Press, 1990).

[23] Kevin Sharpe, *The Personal Rule of Charles I* (New Haven: Yale University Press, 1992) is the definitive study.

[24] In "Antiformalism, Antimonarchism, and Republicanism in Milton's 'Regicide Tracts'" *MP*, 108.4 (May 2011): 509, William Walker argues that "the label, 'regicide tracts,' is a misnomer" and that "in this prose, Milton does not repudiate monarchy."

[25] On *Areopagitica*'s political metaphors of health and disease, see David Norbrook, *Writing the English Republic: Poetry, Rhetoric, and Politics, 1627-1660* (Cambridge: Cambridge University Press, 1999), p. 137.

old kingdom ultimately fail. The poet cannot entirely do away with the traditional monarchical structures, however ruined they might be. The dynastic nation proves impossible to wipe out completely, as does the potential for Restoration, which is activated in 1660.

The identity of England thus remains in flux throughout the century, subject to revision on multifarious fronts. The very name of the nation becomes politically fraught: names begin to proliferate in relation to the multiplying of sometimes contradictory national visions. Drayton's afore-mentioned Great Britain alludes in part to William Camden's *Britannia.* "Great Britain," for Camden and, later, for Drayton, is a geographical entity: one of the two main islands that make up the British Isles. Ireland is the other main island. As a land-formation, Great Britain contains England, Wales, and Scotland. But, "Great Britain," for Drayton, is also a political idea: James I's vision of a unified Scotland and England, a Union that was never, in fact, realized. Drayton's goddess-like Great Britain re-places Queen Elizabeth and Elizabethan England as a dynastic state. As Helgerson notes, Drayton

> seats his figure in a position strongly reminiscent of that assumed by Queen Elizabeth on Saxon's frontispiece, an image that was itself already an adapta-tion of the familiar icon of the Virgin Mary as the queen of heaven. As the cult of Elizabeth had replaced the cult of the Virgin Mary, so the cult of Britain now assumes power in its turn.[26]

However, Drayton's Great Britain is less stable than it appears at first glance. As Claire McEachern observes, we should not assume that the frontispiece is "a synecdoche for the poem's essence."[27] Although greatly favored by Elizabeth I, Drayton was dismissed by James I. *Poly-Olbion* reflects Drayton's ambivalence toward his new monarch: it both celebrates and undermines James's vision of a Greater Britain.

Notably, the most straightforward and outspoken opposition to the Jacobean idea of Union is made in the name of Mother Country: England rather than Britain or Albion/Olbion—as in Drayton's *Poly-Olbion.* On April 25, 1604 Richard Perceval, MP for Richmond, Yorkshire, urged Parliament "the name of our Mother Country to be kept: Our desire natural and honorable—she hathe nurst, bred, and brought us up to be men."[28] Most relevant here are the ways that Perceval's statement places

---

[26] Helgerson, p. 120.

[27] McEachern, p. 167.

[28] Simon Healy, "Perceval, Richard (*c.* 1588–1620," *Oxford Dictionary of National Biog-raphy*, online edn, Oxford University Press, September 2006: http://www.oxforddnb.com/view/article/21914; *Common Journals* (London, 1803), I (1547–1628): 184. See McEach-ern for discussion of Perceval and participants in the debate over the naming of the united England and Scotland, pp. 140–5.

the received trope of England as maternal nurse in service of an imagined English insularity—an England that excludes Scotland. James I appropriates this affecting anti-Unionist equation between nation and nursing by representing his sovereignty as a nursing kingship. I shall address these issues at greater length in Chapters 2 and 3, but nevertheless now wish to underscore the political power and influence of those who remain attached to England as their nursing mother.

In the end, James failed to bring about Union between Scotland and England.[29] Ironically, in 1643, a union temporarily was achieved, but in opposition to the Stuarts. The Scots entered into a formal military-and-religious alliance with the English Parliament against the Stuart monarchy through the Solemn League and Covenant. After the first civil war, the execution of Charles I, and the ascent of the Independents in Parliament, however, the Scots' loyalties shifted to the Royalists. Scotland recognized Charles, Prince of Wales, as King Charles II and joined forces with the Royalists against the English Independents. Negotiations between the Scottish Covenanters and Charles II began in Breda in the Netherlands in March, 1650. The Scots agreed to support Charles II against Parliament and the Independents, but only if the King agreed to take the Oath of the Covenant. To the dismay of the English Royalists, Charles, desperate for allies, accepted. Charles II signed the Treaty of Breda on May 1, 1650, and before arriving in Scotland he took the Covenant, on June 23, 1650. Scotland's shifting alliances newly politicized the distinctions between the terms "England" and "Britain." Especially after 1650, the anti-Stuart opposition favors "England," over "Great Britain," pointedly to exclude Scotland, now allied with Charles II, from its vision of national reformation.[30] (My use of "Great Britain" and "England" reflect these historically specific, political-and-religious shifts.) Tellingly, in 1670 Milton, disappointed by his countrymen's restoration of monarchy, publishes *The History of Britain, That Part Especially Called England*, in which he clearly details the weaknesses of the national character, from top to bottom:

---

[29] A lasting Union between Scotland and England was realized in 1707, when the Parliament of Scotland passed the Union with England Act, and the Parliament of England passed the Union with Scotland Act. These Acts retained their legal currency for almost three centuries, until 1999, when they were modified to allow for the opening of a devolved Scottish Parliament. Scotland nevertheless still remains a constituent country of the United Kingdom. See Christopher A. Whatley, *The Scots and the Union* (Edinburgh: Edinburgh University Press, 2006).

[30] See S.R. Gardiner, *History of the Commonwealth and Protectorate*, vol. 1 (London: Longmans, Green, & Co., 1903), chapter 8, and David Stevenson, *Revolution and Counter-Revolution in Scotland, 1644–51*, 1977, rpt. (Edinburgh: John Donald-Birlinn, 2003), pp. 130–4.

Thus they [members of Parliament] who but of late were extolled as great deliverers, and had a people wholly at their devotion, by so discharging their trust as we see, did not only weaken and unfit themselves to be dispensers of what liberty they pretended, but unfitted also the people, now grown worse and more disordinate, to receive or *digest any liberty at all.* (*CPW*, 5. 449)

Whereas in 1651 and 1653 Milton highlights England in *Pro Populo Anglicano Defensio* and *Defensio Secunda*, his two great *Defence[s] of the* **English** *People* (emphasis added), in 1670 he identifies England as part of Britain, the dissolute Restoration kingdom from which he now feels alienated. As we shall see, the figure of the new nursing mother discursively intervenes in many competing, reformed and traditional, visions of mother country.

Perhaps the clearest reason why interest in maternal nursing intensifies in this unstable period is that mother's milk was thought to have a powerful influence on the formation of identity. Much more than a simple matter of physical nourishment, breast-feeding determined the child's entire being. As Valerie Fildes observes: "Before 1800, the wet nurse (or breast-feeding mother) did not just provide nourishment of the baby; she was believed to transmit to the child, along with her ideas, beliefs, intelligence, diet, and speech, all of her other physical, mental, and emotional qualities. Effectively, she was seen to be reproducing herself: the child *was* the nurse: an extero-gestate foetus."[31] "We may be assured," writes James (Jacques) Guillemeau, "that the Milke (where with the child is nourish'd two years together) hath a power to make children like the Nurses, both in bodie and mind; as the seed of the Parents hath a power to make children like them."[32] The power of nursing to transmit identity is equal to, or greater than, the power of the parents' "seed...to make children like them." In his 1607 account of "The Maners of the Irishry Both of Old and Later Times," William Camden observes that in Ireland the bonds between foster children of the same nurse are much stronger than the kinship or blood-bonds between children of the same biological family:

All those that have been noursed by the same women love one another more deerely, repose greater trust in them, then if they were their naturall whole brethren and sisters, insomuch as in comparison of those and for their sakes they even hate their naturall whole brethren and sisters... [they] break out oftentimes even unto open warre against their said [biological] parents.

Thus, whereas the likeness between nurse and infant can sometimes strengthen the bonds between parents and children, it can also weaken

---

[31] Valerie A. Fildes, *Breasts, Bottles, and Babies: A History of Infant-Feeding* (Edinburgh: University of Edinburgh Press, 1986), p. 99.

[32] James (Jacques) Guillemeau, *The Nursing of Children. Wherein is set downe, the ordering and gouernment of them* from affixed to *Childbirth, or the Happie Deliverie of Women* (London, 1612), sig. Ii4.

natural kinship bonds: "even unto open warre." Nurse milk creates new affective ties among children/subjects who are not related by blood.[33] These affective ties are stronger and more enduring than biological ones. Small wonder, then, that Spenser should find the Anglo-Irish practice of hiring Irish wet nurses so alarming.

The assumption that nursing can bring about a second birth, imprint alternative, anti-customary identities, and foster a spiritual fraternity (based on love) more powerful than a natural one (based on blood) especially enhances the value of breast-feeding for seventeenth-century English Reformers. Their audience is the up-and-coming middling classes, and their anti-wet-nursing rhetoric is directed at these readers: unlike the English aristocracy and nobility, the reformed English mother must nurse her own babies rather than put them out to wet-nurse. This same argument for maternal breast-feeding is reiterated time and again in early modern domestic guidebook literature. This popular middle-class Puritan genre, which first emerges in England at the end of the sixteenth century, makes its case by defending maternal breast-feeding as healthier for children and as the more godly and natural practice for mothers. In the guidebooks, wet-nursing is reduced to a dubious mark of high social status. As the frequent re-printings of Robert Cleaver's and John Dod's *A Godly Form of Household Government*, among other Puritan domestic guidebooks, attest, this Puritan campaign against wet-nursing continues all the way through the seventeenth century and into the eighteenth.[34] Hitherto a calculable form of lower-class labor, breast-feeding in these texts is revised into a mother's highest calling: an incalculable expression of love and duty in response to both God's will and natural law.

Similar views of maternal breast-feeding also can be found in humanist treatises, such as Erasmus's "The New Mother" (*Puerpa*), and aristocratic women's writings, such as Elizabeth Clinton's *The Countess of Lincolnes Nurserie*. In "The New Mother," Eutrapelus appeals to nature when he exhorts Fabulla to nurse her own babies rather than give them to a wet nurse: "But if Nature gave you strength to conceive, undoubtedly it gave you strength to nurse, too."[35] Erasmus's "new mother" is a natural nursing

---

[33] William Camden, *Britannia: Or a Chorographicall description of the flourishing Kingdoms of England, Scotland, and Ireland...* (1607), with an English translation by Philemon Holland. Hypertext edition by Dana F. Sutton (University of California, Irvine, 2004), "The Maners of the Irishry Both of Old and Later Times," Section 5.

[34] Robert Cleaver and John Dod, *A Godly Form of Household Government: For the Ordering of Private Families, according to the Direction of God's Word* (London, 1621); William Gouge, *Of Domesticall Duties. Eight Treatises* (London, 1622). Chilton Latham Powell discusses the print history of Gouge's text in *English Domestic Relations 1487–1653: A Study of Matrimony and Family Life in Theory and Practice As Revealed by the Literature, Law, and History of the Period* (New York: Columbia University Press, 1917), pp. 132–3.

[35] Desiderius Erasmus, "The New Mother," *The Colloquies of Erasmus*, trans. Craig R. Thompson (Chicago: University of Chicago Press, 1965), p. 272.

mother. Elizabeth Clinton invokes both God and nature to make the same point about the naturalness of nursing motherhood: "I will first show that every woman ought to nurse her own child; and second I will endeavor to answer such objection as are used to be cast out against this duty to disgrace the same." "Be not so unnatural to thrust away your own children," she proclaims. Unlike wet-nursing, maternal breast-feeding expresses obedience to and communion with God. The Countess urges other mothers to meditate on these points: "Thinke alwaies that, having the child at your breast and having it in your arms, you have Gods *blessing* there... Thinke againe how your Babe crying for your breast, sucking hartily the milke out of it, and growing by it, is the *Lords own instruction*."[36] These otherwise very different writers—Puritans, humanists, and aristocratic women—thus convey a remarkably similar new concern with maternal breast-feeding.[37]

Reformation interest in the nursing mother can be better understood in comparison and contrast to Catholic veneration of Mary, Mother of Christ, a figure jettisoned by Protestant iconoclasts. The cult of Mary dates to apostolic times: the earliest images of the Virgin Mother were found in the Christian Catacombs of Priscilla in Rome, dating from the third century. Early Byzantine religious art developed several types of Madonna, all emphasizing her role as Christ's mother. One especially relevant variant is the *Theotokos Galaktotrophousa*: the Milk-Giving Madonna. Equally important to my study is the *Glykophilousa Madonna*, the Virgin of Tenderness.[38] As we shall see in later chapters, maternal milk as tenderness is a standard *topos* not only in the domestic guidebooks but, surprisingly, also in Oliver Cromwell's political speeches.

In the fourteenth century, the *Madonna del Latte* dominates especially the decorative arts in Italy.[39] Ambrogio Lorenzetti's *Madonna del Latte* (*c.*1340, Pallazzo Archivescovile, Siena), of the Sienese school, is an especially notable example (see figure 3).

---

[36] Elizabeth Clinton, *The Countess of Lincolnes Nurserie* (Oxford, 1622), Sig. B1r, Sig. D2v.

[37] See Valerie A. Fildes, *Breasts, Bottles, and Babies: A History of Infant-Feeding* (Edinburgh: University of Edinburgh Press, 1986) and *Wet-Nursing: A History from Antiquity to the Present* (Oxford: Basil Blackwell, 1988).Analysis of Puritan attitudes toward breast-feeding can be found in: David Leverenz, *The Language of Puritan Feeling: An Exploration in Literature, Psychology, and Social History* (New Brunswick, NJ: Rutgers University Press, 1980); R.V. Schnucker, "The English Puritans, Delivery and Breast-Feeding," *History of Childhood Quarterly*, 1 (1974): 637–58; and Dorothy McLaren, "Fertility, Infant Mortality, and Breast Feeding in the Seventeenth Century," *Medical History*, 22 (1978): 380–8.

[38] Marilyn Yalom, *A History of the Breast* (NY: Knopf, 1997), chapter 1.

[39] See Caroline Walker Bynum's landmark study, *Jesus as Mother: Studies in the Spirituality of the High Middle Ages* (Berkeley, Los Angeles, and London: University of California Press, 1982) for documentation that the *Madonna del Latte* promotes a pro-maternal breast-feeding agenda.

This tempura-on-wood rendering of the Holy Mother nursing the baby Jesus, who appears to be about three or four years old, is remarkable in that, despite the Madonna's tiny, unrealistic breast, the painting dispenses almost entirely with allegory and symbolism, anticipating the turn in the fifteenth and sixteenth centuries toward greater naturalism. Not unlike Lorenzetti's nursing Madonna, the *Madonna del Latte* (1490, Museum of Fine Arts, Boston) of Bramantino (Bartolomeo Suardi) moves between the allegorical and the natural in its portrayal of the Virgin Mother (see figure 4).

The divinity of Mary's motherhood is allegorized by the halo above her head, but she nevertheless also shows her humanity by nursing the Baby Jesus from her bared left breast, which emerges most unnaturally as a small cone from a slit in her red dress. Although the Baby Jesus is

Fig. 3. Ambroglio Lorenzetti, *Madonna del Latte*, 1330s.

Fig. 4. Bramantino, *Madonna del Latte*, 1490.

swaddled, he is depicted, in the medieval style, as a miniature adult. Facing forward, he stands on two very stout legs with his mother's highly stylized breast in his hand. By the sixteenth century, the shift from allegory to realism is complete. In Correggio's *Virgin and Child with Angel* (1523, Museum of Fine Arts, Budapest), for example, the Madonna offers her naturalistically rendered left breast to a plump and happy Baby Jesus, but he, perhaps already sated, turns away from her breast and toward the equally plump and cherubic Angel (see figure 5).

One of his hands nevertheless firmly holds onto the Holy Mother, even as the other hand reaches toward the Angel. Ironically, just as its naturalistic style is perfected, the *Madonna del Latte* is officially condemned by Pope Paul IV in 1559, after the Council of Trent (1543–63) forbids the depiction of nudity in churches.

As already noted, English Puritan guidebook authors, humanists, and pamphleteers, among other Northern European Reformation writers, also closely relate maternal nursing with nature. However, their idea of

Fig. 5. Corregio, *Virgin and Child with an Angel*, 1523.

the natural, nursing mother is quite different from that of the Italian Renaissance painters for whom the nursing Mary remains divine, even when rendered in a naturalistic style. Protestants, by contrast, remove all overt signs of divinity from the nursing mother. Rather than focus on the unique attributes of the Madonna as the nursing Mother of Christ and as a spiritual mediatrix between Jesus and the people, Reformation writers affiliate Mary with Sarah and Jochobed as a paradigmatic nursing mother: Mary is but one of a long scriptural line of exemplary, natural nursing mothers. For Protestants, this fundamental shift in focus makes the role of the nursing mother "new", as Erasmus emphasizes, despite the long iconic history of this figure. Occluding Roman Catholic iconography of the nursing Madonna, Protestants tend instead to invoke Old Testament imagery of God as a nursing father/mother, bearing the Israelites in his bosom, for example in Isa. 49: 1, 15, in order to underscore the monist-Hebraic attributes of the Reformed nursing mother. This important, but hitherto unexamined Puritan-Hebraic model of the nursing mother is closely analyzed in my study.

The newness of "the new mother" additionally can be attributed to broader cultural changes in attitudes toward children and childcare. In his landmark study, Philippe Ariès argues that an astonishing new interest in and concern for children in their own right—or what he terms "the invention of childhood"—can be clearly detected in France, beginning in the fifteenth and culminating in the seventeenth century.[40] New interest in children as children, rather than as miniature adults, also is detectable in seventeenth- and eighteenth-century England.[41] (As we shall see, the new mother is the key to this new, modern distinction between childhood and adulthood.) Just as the old organic nation begins to unravel, so too does the traditional family. Whereas the old family disciplined unruly children/subjects by force and awe, the new family was to nurture the entrepreneurial and ethical self-direction required of the modern political subject.[42] By lavishing nurture on children, the new family's mothers and fathers could cultivate an enlightened sense of self-worth and ethical/ social responsibility in their offspring. This new combination of self-confidence and social consciousness would prove crucial to their eventual performances as mature, self-regulating citizens of the reformed nation. In the new family, maternal nurture is the key to the making of early modern English citizens.

Narrating experiences vividly remembered long into adulthood, sixteenth- and seventeenth-century personal accounts of wet-nursing and weaning offer important insights into the political and religious contestations that inflect changing early modern conceptions of the family/nation. These accounts also corroborate in subjective terms what we know about the actual conditions of infancy and childhood in the early modern period. Disease, neglect, poor nutrition, feeble limbs, which restricted children's physical movement and mobility, all contributed to the high rate of mortality among infants and children. Moreover, as Janet Adelman

---

[40] Philippe Ariès, *Centuries of Childhood*, trans. Robert Baldick (New York: Vintage, 1962).

[41] Ruth Perry locates the same kind of new interest in children as children, rather than as miniature adults, in seventeenth- and eighteenth-century England. But unlike Ariès, Perry argues that England's "invention of childhood" is inseparable from what she terms "the invention of motherhood," which assigns nursing mothers of all classes a primary role in shaping their children's character and identity. Ruth Perry, "Colonizing the Breast: Sexuality and Maternity in Eighteenth-Century England." *Eighteenth-Century Life* 16 (1992): 194.

[42] Jay Fliegelman discusses these new family values and childcare practices in the context of colonial America, in *Prodigals and Pilgrims: The American Revolution Against Patriarchal Authority, 1750–1800* (Cambridge: Cambridge University Press, 1982).

points out, "a prolonged period of infantile dependency" sparked separation anxieties and the sense of maternal abandonment in adults: children were routinely nursed until the age of two or three.[43] The child's dependency is painfully recollected by the adult as maternal deprivation. Jacques Guillemeau, in *The Happie Nursing of Children*, for example, cites Gauis Gracchi to emphasize the terrible emotional void that wet-nursing creates in children: "I know (Mother) that you bore me nine moneths in your womb, yet that was out of necessitie, because you could do no otherwise; but when I was borne you forsooke me." For Guillemeau, the effects of wet-nursing are so damaging that he associates a mother's refusal to nurse her own child with infanticide: there is "no difference between a woman that refuses to nurse her own child and one that kils her child, as soone as she hath conceived; that shee may not be troubled with bearing it nine months in her wombe. For why may not a woman with as good reason, deny to nourish her childe with her blood, in her wombe, as to deny it her milke being borne?"[44] John Jones's treatise on pediatrics, *The Arte and Science of preseruing Bodie and Soule in all healthe, Wisedome, and Catholike Religion* (1578), dedicated to Elizabeth I, also includes painful reminiscences of nursing and maternal abandonment. Jones recalls that he himself was wet-nursed for three years, much longer, he believes, than is advisable. Not unlike Guillemeau, Jones accuses mothers who employ wet nurses, rather than breast-feed their own children, of "idleness, delicacy, or wantonesse." Conceiving a child is "pleasure" for women, "but to hold it and bear it in her feeble and weak arms, to swaddle it daily on her loving lap, and to give it suck with her own most tender breasts, I affirm with *Tacitus* to be a manifest and undoubted token of absolute kindness and friendship." A child denied the maternal breast will never know "absolute kindness and friendship"; he will be forever bereft and subject to terrible adversity. "For (as Ennius saith) when is a friend tried but in adversity; and who I pray you hath more need than the infant," writes Jones, with, it would seem, himself and all other wet-nursed babes in mind.[45] As Boyd Berry observes, Jones's argument here "puts us (and certainly Jones)

---

[43] Janet Adelman, *Suffocating Mothers: Fantasies of Maternal Origin in Shakespeare's Plays, Hamlet to The Tempest* (New York: Routledge, 1992), p. 5.

[44] James (Jacques) Guillemeau, *The Nursing of Children*, the quotations are from the "Preface" (I.i.2) and "The Preface to Ladies." Guillemeau refers to Gauis Gracchus rather than to his brother, Tiberius Gracchus, the ill-fated reformed, since only Gauis' writings survive from antiquity. See H. Malcovati, ed., *Oratorum Romanum Framenta Libera Rei Publicae* (4th edition, Turin, 1976–9), pp. 174–98.

[45] John Jones, *The Arte and Science of preseruing Bodie and Soule in all healthe, Wisedome, and Catholike Religion* (London, 1578), p. 30. Cited in Boyd M. Berry, "The First English Pediatricians and Tudor Attitudes Toward Childhood," *Journal of the History of Ideas*, 35.4 (October–December, 1974): 576.

in the position precisely of a hungry infant, depraved through sex and sin, experiencing infantile adversity."[46]

But, while painful and injurious to the private individual, the deprivation and dependency experienced by children in the traditional family (the family without the nursing mother) nevertheless could prove advantageous to the governing social structures and political institutions of monarchy. Injured, dependent children would turn into docile, fearful adults and, by extension, obedient Tudor–Stuart subjects. By contrast, children in the new family/nation were to drink in a coherent sense of self and individual autonomy with their mother's milk. Rather than the disruptive and wounding experience alluded to by Guillemeau and Jones, maternal nursing and the process of weaning would be experienced as wholesome, natural, and holy, a point that Puritan guidebook writers go to great lengths to underscore. The new family is designed to spare children the painfulness of a separation that could be neither forgiven nor forgotten, and which created psychic wounds that could never be healed. Integral and intact, rather than broken and dependent, reformed children promised to mature into engaged, ethical adults: future, self-directed citizens of the new, post-dynastic nation. Just as it represents a means of securing national identity, so maternal nursing also comes to be seen as a natural mechanism for repairing the broken spirit and damaged character of the English people.

This is a project of no small import. The need to find new ways to restore the physical and spiritual health of the English people acquires increasing importance and urgency to the Reformers, many of whom link political liberty to the cultivation of a strong national character. A good number of these national-character-building programs centrally address the home, the family, and, especially, the breeding, both biological and social, of children. The popular republican, John Streater, argues that "the increasing of heroickness in the breast of the people" is indispensable to winning freedom from censorship, among other forms of political oppression. For Streater, creating a heroic people could be best achieved by selective breeding, which would allow only the strongest children to be produced.[47] Whereas Streater advocates eugenics, John Amos Comenius, the influential educational reformer, espouses universal education or

---

[46] Boyd M. Berry, "The First English Pediatricians and Tudor Attitudes Toward Childhood": 576.

[47] John Streater, *Observations, Historical, Political and Philosophical, upon Aristotle's first Book of Political Government* (London, 1654): no. 6, p. 44. See Nigel Smith, "Popular Republicanism in the 1650s: John Streater's 'Heroick Mechanicks'," in *Milton and Republicanism*, ed. David Armitage, Arman Himy, and Quentin Skinner (Cambridge: Cambridge University Press, 1995), pp. 137–55, especially p. 143.

"pansophy" as the best means to strengthen the English character and to produce a strong and virtuous new England. For Comenius, the young provide the best seeds for growing the nation's new character. He identifies childhood as the ideal moment to breed social change and to naturally nurture civic virtue and heroic valor through a reformed process of education. His metaphors of planting and gardening underscore this point:

> You have seen that there is no better time than early youth to fight against the evils human are prone to as a breed, no better ways to plant a tree destined for eternity than to plant it and train it up while it is young, no better time to raise Zion on the site of Babylon than when the living building-stones of God, the young, are easily hewn shaped, polished, and fitted for the heavenly edifice.[48]

In *Of Education*, a tract dedicated to Samuel Hartlib, an English disciple and translator of Comenius, Milton similarly identifies childhood as the best time to renew and strengthen the English character. Unlike Streater and Comenius, however, Milton relies upon the Galenic principle that a weak or unruly character can be corrected by proper nurture and diet. This humoral remedy is crucial to the poet's educational goal: to transform unformed, undisciplined youth into free-thinking, patriotic citizens and heroic, public servants of the reformed nation. Praising other authors (left unnamed, possibly, Erasmus, Gouge, among others), who focus on the nurture of children from the cradle, Milton acknowledges that although he himself is unable to address the issue, he nevertheless believes that the proper care and nurture of infants and young children are important to their future maturation into English patriots. I shall spend much more time on Milton's tract in Chapter 3, but for now I wish to underscore that, like many of the writings I shall investigate, *Of Education* provides crucial textual evidence of how the new ideology of nurture is interpolated within larger Reformist programs for remaking the English nation.

It is worth repeating that the new nursing mother remains a highly contested figure throughout the period, capable of serving the interests of the old nation as well as the new one. The nursing mother is celebrated in writings that wish to revive and reinvigorate the old nation's natural bonds of blood and genealogy, even though in other texts, she innovatively generates the bonds of love and spiritual fraternity that organize the new affective community of the reformed nation. As I shall demonstrate in Chapter 3, James I, Charles I, Cromwell, and Milton all recognize the new primacy of nurture in creating national cohesion.

---

[48] John Amos Comenius. *The Great Didactic of John Amos Comenius*, trans. M.W. Keating (New York: Russell and Russell, 1967), p. 18.

Further complicating the contested figure of the nursing mother is that, in addition to lending a satisfying sense of union to the nation, whether old or new, she also smoothes over the ambivalences that are intrinsic to the emerging modern nation, thus strengthening its conceptual foundations. Innovative theories from the last three decades highlight the modern nation's contradictory aspects. In Tom Nairn's well-known formulation, what distinguishes the old nation from the new is the modern nation's Janus face: that it is bellicose and humanitarian, expansionist and bounded, grounded in law and reason but capable of arousing intensely passionate patriotic feelings, all at the same time.[49] Robert J.C. Young observes that the modern nation is an "impossible hybrid": it "exhibits the characteristics of modern culture, namely that it is conflictual and divided against itself." For Young, the modern nation and nationalism are difficult to define because their meanings "are generated by the dialectic that Adorno has called the 'torn halves' of a culture that does not add up."[50] As we shall see, the figure of the new mother plays an important but surprisingly under-analyzed role as a private–public conduit between the "torn halves" of the new nation. She lends symbolic unity and stability to the emerging nation's foundational contradictions and liminality: its foundational neither-here-nor-there place and time between the outmoded structures of the past and the unrealized social promise of the future.

William Marshall's engraved frontispiece to Richard Brathwaite's conduct book, *The English Gentleman and English Gentlewoman*, can help us to picture the otherwise almost indiscernible and hitherto unrecognized interplay between the comprehensive, yet contested figure of the new mother and the doubleness of the emerging nation (see figure 6).[51]

Like Streater, Comenius, and Milton, Brathwaite offers a program for building the character of the English people, although his class-based emphasis is on the moral acquisition of English gentility rather than the natural breeding or training up of national heroes. Brathwaite's frontispiece also can also be usefully compared with Drayton's. Like Drayton, Brathwaite draws on the satisfying sense of unity and coherence projected by the figure of the nursing mother. But whereas Drayton's Great Britain

[49] Tom Nairn, *The Break-Up of Britain: Crisis and Neo-Nationalism*, 2nd ed. (London: Verso, 1981), pp. 348–9.

[50] Robert J.C. Young, "The Overwritten Underwritten: Nationalism and its Doubles in Postcolonial Theory," *The Silent Word: Textual Meaning and the Underwritten*, ed. Robert Young, Ban Kah Choon, and Robbie B.H. Gobb (Singapore: University of Singapore Press, 1998), pp. 1–3.

[51] Richard Brathwaite, *The English Gentleman and English Gentlewoman* (Wing/B4262, 1641), Frontispiece.

Fig. 6. Richard Braithwaite, frontispiece, *The English Gentleman and The English Gentlewoman*, 1641.

is identified with the land and its natural resources (its people and produce), Brathwaite's national mother is associated with an abstract perception of the English character as comprised of various intangible qualities:
honor, moderation, and decency, among others. Drayton's goddess-
mother serenely occupies the center of the frontispiece; Braithwaite's maternal figure is much smaller and positioned at the top-center, above the
word "English" in the title. As Jacqueline T. Miller observes, she is "the
presiding deity of the text," but she is also less fleshy and earth-bound
than Drayton's maternal goddess. Unlike Drayton's goddess, Brathwaite's
"presiding deity" is seated, and, instead of a cornucopia of fruits at her
breast, she has an open book (a prayer-book or Bible, perhaps) in her lap,
where one might expect to find a baby.[52] Both her breasts are bared, and
she holds each nipple between her second and third fingers: she is express-
ing milk. Her breast milk forms two distinct streams: one stream flows
into the manly virtues to her right; the other flows into the womanly
virtues to her left. Brathwaite depicts the English character as written in
breast-milk not ink, a substitution with clearly gendered implications: the
breast replaces the pen as the authorial implement of choice.

In Brathwaite's frontispiece, the new mother supersedes the old father;
abstract virtues replace blood and soil as the defining measures of English
gentility.[53] The old organic standards of status are exclusive: they limit
access to the upper class by tying aristocratic status to elite bloodlines and
genealogies. The new abstract standards of class are inclusive: rather than
a particular biological pedigree, universal Christian standards of morality
determine entry to the upper class. Not only does aristocratic status
become accessible to a wider sector of the population (a moral elite), but
also the middling class's enhanced economic power—including, the
power to purchase titles and lands which hitherto could only be inherited—gains the moral currency of Christian virtue. Printed guides such as
Brathwaite's conduct book help to put these new principles into practice
by tutoring their middle-class readers on how to acquire the specific virtues and manners they will need to fashion themselves as English gentle-
men and gentlewomen, as defined under the reformed system.

In addition to changing perceptions of status, class, honor, and nobility, Brathwaite's abstraction of the English character from the land and/or
the body of the hereditary monarch reflects the period's shifting paradigms of the nation. Land and ruler are nowhere to be found on
Brathwaite's frontispiece, unlike Drayton's, which, as we have seen, pushes

---

[52] Jacqueline T. Miller, "Mother Tongues: Language and Lactation in Early Modern
Literature," *ELR*, 27 (1997): 178.
[53] Michael McKeon, *The Origins of the English Novel 1600–1740* (Baltimore: Johns
Hopkins University Press, 1987), pp. 162–6.

the old, anxious male rulers to the peripheries and makes the mother-land the serene, governing center of a greater Britannia. At the beginning of the century, the mother-land can be imagined as peaceful, procreative, self-sufficient, and map-able; in 1641, however, when the nation is on the verge of civil war, it is mutable, embattled, un-chartable, and, for Brathwaite, abstract and allegorical as well. At this tumultuous historical moment, neither ruler nor land nor map proves either viable or desirable as a measure of national unity. Whereas Drayton can depict the nursing mother-land as the sole progenitor and nourisher of the nation and its people, Brathwaite needs to push the land—but not the figure of nursing mother—beyond the purview of the reader in order to establish the new abstract moral ground, in which, for him, England—as opposed to Drayton's landed Great Britain—must now be rooted.

Rather than spring autochthonously from England's mother-earth and gain sustenance at her breast, the English character is comprised entirely of virtues gleaned from the printed page of Brathwaite's book. Text replaces the land as the source of national identity. At the same time, however, the frontispiece attributes the character-building powers of Brathwaite's book to the presiding figure of the nursing mother. Brathwaite seems to want to define Englishness in both ways: as abstract and textual, and as embodied and natural. As she does elsewhere, the nursing mother both exhibits and covers up underlying ambivalences, but the conflicts here are rather different from those smoothed over by Drayton's nursing goddess. Brathwaite's English mother is inspired to nurse by the pious book-as-baby that she holds in her lap. But, she also expresses this book as breast-milk, which, as it flows down both sides of the title-page, inscribes the new abstract English character—and also naturalizes it. Brathwaite's nursing mother allows Englishness to be organic and symbolic, corporeal and incorporeal, all at the same time. These are the very same paradoxes which, as Nairn, Young, and others demonstrate, constitute the conceptual foundations of the emerging modern nation. Whether by accident or design, Brathwaite's frontispiece puts the new mother and the new nation on the same page. As we shall see in my last two chapters, Milton also conflates (and separates) the contested figure of the nurturing mother with the doubleness of the modern nation, in clearly motivated ways.

II

My book opens with a broad survey of the some of the diverse cultural arenas in which the figure of the new mother and changing paradigms of Englishness most closely overlap. In Chapter 1, I begin by examining constructions of the new nurturing mother in Puritan domestic guide-books, in such texts as the aforementioned Robert Cleaver and John Dod's *A Godly Form of Household Government* and William Gouge's *Of Domesti-call Duties*. In this popular, middle-class genre, the nursing mother is rep-resented as having a governing influence on the formation of identity. The guidebooks are first printed at the end of the sixteenth and beginning of the seventeenth centuries, or at the precise moment when England's sense of its insular identity begins to erode. On the one hand, England begins to assert itself as an imperial and mercantile power and worthy competi-tor to Spain, Portugal, France, and Holland, and, on the other, mounting religious and political conflicts threaten to break down customary meas-ures of national unity from within. England's closer contact with foreign cultures makes the nation's anxieties about its vulnerable insular borders even more acute. Joseph Hall's disoriented response in *A History of the New World* to the new flow of information about China is an excellent case in point: "there is men, perhaps more ciuill then wee are, who ever expected such wit, wuch gouernment in China? such arts, such of all cun-ning? we though learning had dewlt in our corner of the world: they laugh at vs for it, and well may."[54] Early modern English perceptions of China were thoroughly mediated by national, political, and religious dis-putes. I have written elsewhere about how the Jesuits and the Stuarts were the first Europeans to assume propriety over China in intellectual terms.[55] China is in the Far East, but "China" is also located conceptually in the Stuart court and linked to Roman Catholicism. Puritan guidebook au-thors like Gouge, Cleaver, and Dod attempt to remedy this kind of glo-bal–national dislocation by rooting English identity in neither the map-able land (is it China and/or England?) nor in the Sinophilic Stuart court, but instead in the pure and unbreakable natural-and-godly bond between the new nursing English mother and child. As already noted, the new Puritan-Hebraic model of the nursing mother emphasizes the same natural-divine-national character of this primal bond.

The naturalness of the nursing English mother also enables nurture to be assessed and managed in scientific terms as well as scriptural ones. The guidebooks draw not only on both Testaments but also on both the old Galenic science and the new science of obstetrics (which emerges as a discipline in England for the first time in the early seventeenth century)

---

[54] Joseph Hall, *The Discovery of the New World or a Description of the South Indies. Hith-erto Unknowne* (London: 1609), sig. A4r.

[55] "'The people of Asia and with them the Jews': Israel, Asia, and England in Milton's Writings," *Milton and the Jews*, ed. Douglas Brooks (Cambridge: Cambridge University Press, 2008), pp. 151–74.

to prove that maternal nurture is indispensable to the nation's social and spiritual health and well-being. I offer a close reading of William Harvey's *Anatomical Excercitations, Concerning the Generation of Living Creatures* to demonstrate how early obstetrical theories inform, and are implicated in, the new ideology of nurture.[56] I conclude my first chapter by demonstrating that, in addition to domestic guides and medical texts, maternal nursing is a recurrent trope in Puritan sermons. Puritan ministers routinely represent themselves as the "breasts of God," nursing their congregants with the milk of the Word.

In Chapter 2, I examine three texts that span the Jacobean period: two gender-debate pamphlets, *A Pitiliess Mother* and *Hic Mulier*; and Shakespeare's *Macbeth*. *Macbeth* may seem a predictable choice, given Lady Macbeth's oft-cited attempt to demonstrate male mettle through her presumptive act of ripping the babe from her breast and then dashing out its brains. Surprisingly, however, Shakespeare's play has not yet been read in relation to the Jacobean debate about women's worth. In *Macbeth*, as in the pamphlets, the new mother represents the natural means for consolidating diverse regions, classes, and populations into a new social whole. In the gender debate, however, the nurturing mother also serves as the natural organizing principle for an emergent form of affective nationalism. In Shakespeare's play, she is a much more ambivalent, unnatural and natural, figure, one that points both toward the new abstract nation and away from it as well. In the end, I argue, the new mother is both perfected and subsumed by the magical, remedial figure of the old hereditary patriarch: Edward the Confessor. King, healer, and nourisher, the English king restores moral and social health. He is both mother and father: perfect, whole, and absolute. Yet, he never appears on stage. He remains in the wings: a sublime, unrepresentable, androgyne, but also perhaps no longer completely viable as a monarchical player in the new disenchanted era.

Taking *Macbeth*'s androgynous English king as a point of departure, my third chapter focuses on the figure of the nursing father. This venerable trope first appears in the Hebrew Bible ("Kings shall be your nursing fathers" [Isaiah 49:23])[57] and then is adopted by Christianity, in the widely circulated medieval image of Jesus as a nurse.[58] As in Shakespeare's play, the sacred figure of the nursing father is converted into a secular emblem of male political power. This chapter compares Stuart

---

[56] William Harvey, *Anatomical Excercitations, Concerning the Generation of Living Creatures: To Which are added Particular Discourses, of Births, and of Conceptions, &c.* (London, 1653).

[57] *The Holy Bible* (London, 1611).

[58] See Caroline Walker Bynum, *Jesus as Mother: Studies in the Spirituality of the High Middle Ages* (Berkeley, Los Angeles, and London: University of California Press, 1982).

and anti-Stuart incarnations of the nursing father in *Basilikon Doron* and *Eikon Basilikon*, on the one hand, and, on the other hand, Cromwell's public speeches and Milton's *Of Education* and *Areopagitica*. In Milton's tracts, paternal nursing is associated with the new educator and new legislator. Unlike the Stuarts, however, Milton in these two tracts does away with all imagery, scriptural or classical, that concretely or naturalistically conjures up the nursing breast. An iconoclast and anti-monarchist, Milton absents the Hebraic/Stuart image of the king as a nursing father with "nourish-milk" and, instead, evokes only the pure abstractions that this image had hitherto embodied: nurture, love, and charity. When read alongside the Stuarts' more robust figures, however, Milton's abstractions distinctly imply their more concrete antecedents. Yet, whether absented and abstract or undeniably present and concrete, the nursing father in all of these very different texts does a great deal of cultural work: he shores up the absolute, male–female wholeness and *sui generis* nature of the old king and kingdom, masculinizes the toleration and liberality of the new nation and its leaders, and puts a loving, maternal face on tyranny, conquest, and oppression.

My last two chapters investigate the different ways that nurture and the nation intertwine in *Paradise Lost* and *Samson Agonistes*. As noted above, Milton takes the shift from the corporealist paradigm of the dynastic nation to the abstract or imagined community of the modern Janus-faced nation and maps it onto the fraught transition from the Old Testament to the New Testament. The complex relations between the two Testaments and between the Hebraic/Judaic and the Christian—and as we shall see between Adam and Eve—are crucial to my analysis of Milton's changing conceptual reformation of the nation. Although in his prose tracts, Milton explicitly identifies England as the new Israel, in his later poetry, he does not confine the new nation within territorial boundaries. Rather, the new nation is all-inclusive (or, more specifically, it is exclusively inclusive, a point to which I shall return): it encompasses all godly peoples. Milton's abstract, universal nation is a reformed, one-world Protestant community of the spirit. The ambiguous imperial/anti-imperial implications of Milton's paradigm of universal nationhood shall also be subject to critique. In *Paradise Lost*, the new mother is Eve as "second Eve" (5. 349; 10.183); she typologically foreshadows the nurturing Virgin Mother of the promised seed, without explicitly alluding to Mary, a figure jettisoned by Protestantism. "Second Eve" (5.349; 10.183) smoothly fills in the abrupt and painful gap between the end of the old dispensation and the beginning of the new one. In so doing, she also consolidates the ambivalences that are intrinsic to the emerging nation—a nation that Milton universalizes in his epic.

The importance of Eve's consolidating future role as spiritual mother and nurturer is further clarified when contrasted with Adam's diminished status as a biological father after the Fall. Whereas Eve projects a satisfying sense of maternal wholeness onto the new nation's Janus-face, Adam, with his cursed, proliferating progeny, is associated not only with the patriarchs of the Old Testament and their long lines of descendants, but also with the old organic nation and the natural potency of kingship, with its never-ending dynastic line. In the end, Adam learns from Michael how to read "in the spirit," and embraces the Son as his savior: "Taught by his example whom I now/Acknowledge my redeemer ever blessed" (12.572–73). Yet, as I shall demonstrate, despite his transformative acknowledgement of the Son as Redeemer, Adam does not entirely go forward toward the redemptive anti-patriarchal future with Eve: part of him is left behind. Because he cannot completely part with that part of himself that refuses to break "The bond of nature" (9.956), Adam can only partially enjoy the spiritual promise of the redeemed universal-national future. In Milton's imagined Reformation future, the biological father is diminished, whereas the Virgin Mother and her spiritual son gain ascendancy. This is why the poet identifies the Son Incarnate with a non-biological, exclusively maternal line: "Jesus, son of Mary, second Eve" (10.183). In theological terms, the shift from biological father to spiritual mother, and from natural to spiritual progeny, parallels the shift from the restricted food and specific dietary prohibitions of Mosaic Law to universal milk, a commonplace trope for the gospels. The shift from the old organic nation to the new spiritual and affective one goes hand in hand with the shift from the old chosen nation (the biblical Israel) to the new Israel of the faithful, the world over.

Whereas the new mother in *Paradise Lost* plays a pivotal role, in *Samson Agonistes*, she is morally bankrupt, devoid of significance and worth. In *Samson*, Milton demonstrates that the new mother is not new at all. Although the hero has two maternal "nurses," Manoa and Dalila, neither one can provide him with the nurture he needs to revive his prophesied identity as Israel's redeemer. Manoa offers his son an Old Testament (and hence obsolete) paradigm of the self and the community. Dalila is the perversely alluring object of Samson's idolatry and his carnal desire for "Bondage with ease" (*SA*, 271). God is the only "mother" whose nurture is truly generative and restorative. Not unlike other reformist nursing newborns, Samson, as God's nursling, is born a second time in spirit at the end: his new anti-customary heroic identity is made in the nurturing image of the deity. For Milton, God's nurturing powers find their fullest expression in the incarnated Son. As an Old Testament type of Jesus, the reformed Samson fulfills his prophesied identity when he severs his relations with his nurturing mothers (Manoa and Dalila) and turns exclusively to God for spiritual sustenance. At the same time, Samson's restoration is imperfect: his hyper-masculine strength and his brutal acts of destruction at the Dagonalia fail to deliver Israel from Philistine bondage. Samson fails in fact and triumphs in spirit at the end, all at once. He is both one of the failed, false deliverers from *The History of Britain* ("they who but of late were extolled as great deliverers, and had a people wholly at their devotion, by so discharging their trust as we see, did not only weaken and unfit themselves to be dispensers of what liberty they pretended") and, simultaneously, a typological shadow of the truly great deliverer of *Paradise Regained*, published in the same volume as *Samson Agonistes*: "Hail Son of the most high, heir of both worlds/Queller of Satan, on thy glorious work/Now enter, and begin to save mankind" (*PR*, 4.633–5).

In sum, all five chapters try to determine precisely what is at stake for the early modern English nation in the contested figure of the new nurturing mother. The stakes are high indeed: changing ideals of motherhood challenge, subvert, and redefine some of the most intimate of all human relationships: between husbands and wives, and between parents and children. Because the new nurturing mother cuts so deeply into the delicate fabric of the individual and collective psyche, she provides an especially affecting register of the joy and pain, gain and loss, hope and disappointment of Reformation and its sweeping cultural, religious, and political changes—changes that shake the nation at its very core. The revaluation of the nurturing mother (and the devaluation of this revalued figure in *Samson*) opens a new window on the inner recesses of the English family/nation at an extremely fraught historical juncture: after the old organic dynastic nation begins to unravel, but before the new abstract national community is fully conceivable. As most clearly delineated in Milton's prose and poetry, this fraught moment is mapped onto the hyphen that separates the Hebrew Bible from the New Testament in the Judeo-Christian tradition. As already noted, in Roman Catholicism, the Holy Mother tenderly mediates between old and new, erasing conflict and affirming continuity. Protestantism's new mother as natural and/or spiritual nurse stands in for the mediating Holy Mother. She is the new natural–spiritual conduit through which the old era and old nation (organized by biological bonds) gently gives way to the new era and new nation (organized by spiritual bonds). The reformed mother as natural/spiritual nurturer thus plays a hitherto unrecognized, but crucial role in the conceptual paradigm shift that the English nation undergoes: from a dynastic kingdom organized, not unlike the biblical kingdom of Israel, on bloodline, birthright, and kinship bonds to a modern nation in which spiritual bonds supersede biological ones, as in the new spiritual Israel of the Pauline letters. However, despite its emphasis on unity of spirit, the new nation is in fact rooted in its irresolvable tensions between body and spirit, exclusion and inclusion, the Hebraic/Judaic and the Christian— hence its "Janus face." Despite her seductive promise of satisfying spiritual wholeness, the new mother, at best, covers up irreconcilable conflicts that are foundational to the reformed nation. In its most ambitious scope, then, my book strives to illuminate the obscure nexus among the revaluation of maternal nurture, the ambiguous relations between the Hebraic/Judaic and the Christian after the Reformation, and the highly contested idea of the nation in seventeenth-century England.

1

Nursing Mothers and National Identity

In *A Godly Forme of Household Government*, Robert Cleaver and John Dod urge their readers to recognize maternal breast-feeding as a mandate of both God and nature. "Amongst the particular duties that a Christian wife ought to perform in her family, this is one," they proclaim, "namely, that she nurse her owne children: which to omit, and to put them foorth to nursing, is both against the law of nature and also against the will of God."[1] William Gouge, in *Of Domesticall Duties*, celebrates breast-feeding as the highest expression of Christian motherly love: "How can a mother better express her love to her young babe then by letting it suck of her owne breasts?"[2] In *The Doctrine of Superiority*, Robert Pricke similarly exhorts mothers to "nourish [their own children] and [be] most tenderly affected toward them"; they also must recognize that breast-feeding is the consummate expression of motherly love and obedience to God's will and natural law: mothers "shuld minister fit nourishment unto the infant & so set forth the glorie of God."[3]

Domestic guidebooks like those of Cleaver, Dod, Gouge, and Pricke provide important insights into the cultural conflicts and social ambitions shaping the reformation of motherhood in early modern England. These popular texts were frequently reprinted throughout the sixteenth and seventeenth centuries: William Gouge's *Of Domesticall Duties* went through three editions; Cleaver's and Dod's *A Godly Forme of Household Government* went through nine.[4] These texts retained their popularity well into the 1650s and beyond.

How representative, however, are they in fact? Not very, it might seem at first glance. Some guidebooks such as James (Jacques) Guillemeau's *The*

---

[1] Robert Cleaver and John Dod, *A Godly Form of Household Government: For the Ordering of Private Families, according to the Direction of God's Word* (London: R. Field, 1621), STC (2nd ed.) / 5387.5.

[2] Gouge, p. 518.

[3] Robert Pricke, *The Doctrine of Superioritie* (London, 1609), sig B3r, Section K.

[4] Chilton Latham Powell discusses the print history of Gouge's text in *English Domestic Relations 1487–1653: A Study of Matrimony and Family Life in Theory and Practice as Revealed by the Literature, Law, and History of the Period* (New York: Columbia University Press, 1917), pp. 132–3.

Nursing of Children are translations of continental texts.[5] However, most domestic guidebook authors were Englishmen; many were Puritan divines as well. Writing out of a shared belief system, the guidebook writers closely echo one another. With their strong Puritan sympathies, they reflect the common views of a small subgroup and not a comprehensive cultural vantage point. The guidebooks also target a very specific middle-class readership. Their primary aim is to exhort and persuade rather than to document lived experience.[6] These texts had little impact on actual behavior. Despite the guidebooks' persuasive arguments that maternal breast-feeding is both morally superior to and more natural and medically beneficial than wet-nursing, wet-nursing nevertheless remains the norm for a wide range of social classes. Not only did aristocrats and gentry wet-nurse their babies, but the upwardly mobile wives of lawyers, doctors, merchants, and so forth (the middle-class female audience targeted by Puritan guidebook writers) also regularly hired wet nurses.

But while they appear to be doctrinaire, the guidebooks in fact are more ambivalent than they seem at first glance. As Heather Dubrow observes, these texts are riddled by the same moral and epistemological uncertainties that (dis)organize post-Reformation belief-systems, both sacred and secular. They are more representative of early modern culture-at-large than their rather narrow range of assumptions would appear to suggest.[7] Guidebook writers assign both sacred and scientific measures of meaning and value to the figure of the nursing mother. Moreover, their science draws both on traditional Galenic theories of the four humors and on new empirical paradigms of the human body. Obstetrics, as we shall soon see, emerges as a modern scientific discipline for the first time in England during the seventeenth century.

On the one hand, then, the guidebooks define maternal breast-feeding as a biological impulse, which if inhibited, would pervert the natural order of things, as is the case, they maintain, with wet-nursing. They also express genuine interest in the physical health of mothers and infants. After offering twenty-three reasons why women should nurse their own babies, Gouge concludes that children breast-fed by their own mothers

---

[5] James Guillemeau, *The Nursing of Babies,* affixed to *Child-birth, or the Happy Deliverie of Women* (London, 1612).

[6] On the disparities between the guidebooks' idealized perceptions of the English household and actual domestic practice, see Margaret J.M. Ezell, *The Patriarch's Wife: Literary Evidence and the History of the Family* (Chapel Hill: University of North Carolina Press, 1987), pp. 161–3, and Steven Ozment, *When Fathers Ruled: Family Life in Reformation Europe* (Cambridge: Harvard University Press, 1983), pp. 55, 70.

[7] Heather Dubrow documents the guidebooks' interchangeability and their cultural exemplarity in *A Happier Eden: The Politics of Marriage in the Stuart Epithalamium* (Ithaca: Cornell University Press, 1990), pp. 10–20. I draw on her insights here.

"prosper best. Mothers are most tender over them, and cannot indure to let them lie crying out, without taking them up and stilling them, as nurses will let them crie and crie again."[8] Gouge urges women of all classes to appreciate the medical benefits and natural wholesomeness of maternal breast-feeding.

As just noted, the guidebooks' medical recommendations draw upon Galen's humoral model of the human body as porous and absorbent. Helkiah Crooke, physician to King James I, maintains: "For the matter of man's body is soft, pliable, and temperate...for man is the moyestest and most sanguine of all Creatures."[9] Despite its tight grip on early modern culture, Galenism nevertheless starts to unravel in the seventeenth century, when new empirically based accounts of biology begin to prevail over the older humoral paradigms, for which no empirical evidence could be found. Although they continue to imagine the human body in humoral terms, guidebook authors, time and again, challenge one of humoralism's governing assumptions: that mothers' milk was formed from menstrual blood and therefore toxic, especially in the period immediately after childbirth. This belief proved detrimental to infant health. "The number of nurse children that die every yeare is very great," writes Gouge.[10] In 1671 Jane Sharp's comments on the medical benefits and deficits of breast-milk similarly combine the old humoral science and the new empirical one. Although she endorses the Galenic theory that menstrual blood "hath strong qualities indeed, when it is mixed with ill humors," she also claims, based upon her many observations of nursing mothers and children, that "were the blood venomous it self, it could not remain a full month in the womans body, and not hurt her; nor yet the Infant, after conception."[11]

On the other hand, however, the guidebooks proclaim that the true value of maternal breast-feeding can be measured *only* in spiritual terms: that its worth is neither quantifiable nor scientifically verifiable. Hitherto a calculable mode of lower-class labor, nursing is refined into a mother's "speciall calling." It is a vocational impulse, governing all mothers regardless of external and embodied measures of worth and status. As Gouge maintains, "No outward business appertaining to a mother can be more acceptable to God then the nursing of her childe."[12] A strong, natural biological urge, on the one hand, breast-feeding, on the other hand, expresses God's spiritual feeding hand. Guidebook representations of the nursing

---

[8] William Gouge, *Of Domesticall Duties*, pp. 286–9.
[9] Helkiah Crooke, *Microcosmographia: A Description of the Body of Man* (London, 1615; STC 6062.2), p. 5.
[10] Gouge, p. 518.
[11] Jane Sharp, *The Midwives Book* (London, 1671; Wing S2969B), p. 289.
[12] Gouge, p. 286.

mother thus oscillate between the concrete and the abstract, science (both old and new) and sacred text, and the natural and the spiritual, all at the same time: they never quite add up.

Following Dubrow, I read the guidebooks' contradictory revaluations of maternal breast-feeding as representative of the broader conflicts that shape early modern English culture as a whole.[13] However, I also wish to take seriously the ways that the guidebooks promote very specific class interests by implementing new, proto-middle-class measures of value and significance. These texts contribute to the early modern cultural translation of traditional blood-and-soil measures of noble status into abstract, moral, and emerging class terms, in response to the social mobility fueled by expanded commerce and mercantilism in late sixteenth- and seventeenth-century England. I am especially interested in this shift from organic to abstract measures of moral-and-class value because it generates the same kind of conceptual contradictions that are foundational to the modern nation, organized by abstract social bonds, but nevertheless still rooted in the body. I contend that gender is the key to understanding these contradictions. Although sometimes obscured in recent studies of the modern nation, the maternal plays an important role in shaping and/ or consolidating early modern culture's unstable paradigms of status, class, and national identity. By emphasizing the central role that maternal breast-feeding plays in the transmission of identity, the guidebooks establish the equally embodied-and-disembodied figure of the new nursing mother as the connecting thread between the shifting organic-and-abstract middle-class measures of social value and national identity.

A review of the customary value and meaning accorded to the figure of the nursing mother can help to highlight the guidebooks' new sacred-scientific assessments. In the early Middle Ages, Mary played a minor role in the Christian story. However, with the rise of a more emotional Christianity, associated with the preaching of Bernard de Clairvaux in the twelfth century, Mary became the cultic focus of popular piety. Unlike the scholastics who took a rationalist approach to understanding divinity, Bernard preached an immediate faith in which the Virgin Mary was the intercessor between the people and Jesus. Bernard's special devotion to the Virgin is captured in Alfanso Cano's seventeenth-century painting *Bernard Clairvaux Lactactio*, in which the Holy Mother, with the Christ child in one arm, expresses milk from her right breast straight into Bernard's mouth (1650, Museo de Prado, Madrid) (see figure 7).

---

[13] Michael McKeon notes that Protestant attempts to accommodate the sacred to a profane world, along with the ensuing tensions between progress and regress, gain and loss, "haun[t] seventeenth-century England on every level of thought and activity," p. 65.

From the medieval to the early modern period, Mary's milk was considered the most holy and most miraculous of fluids, next to the blood of Christ.[14] As Carolyn Bynum documents, from the twelfth century to the fourteenth century and beyond, numerous images from art and literature contain explicit parallels between the blood that pours from Christ's wounds and the milk that flows from Mary's breasts.[15]

The commonplace allegory of Charity as a nursing mother also comes out of a long exegetical tradition.[16] The ninth-century theologian Walafrid

Fig. 7. Alonso Cano, *The Vision of Bernard Clairvaux*, 1650.

[14] See Marilyn Yalom, *A History of the Breast* (New York: Knopf, 1997), esp. pp. 43–5.
[15] Carolyn Bynum, *Holy Feast and Holy Fast: The Religious Significance of Food to Medieval Women* (Berkeley: University of California Press, 1987), pp. 269–76; and *Jesus as Mother: Studies in the Spirituality of the High Middle Ages* (Berkeley: University of California Press, 1982).
[16] Marina Warner, *Alone of All Her Sex: The Myth and Cult of the Virgin Mary* (New York: Vintage, 2000), p. 194.

Strabo was among the first to allegorize the breasts of the bride of the Song of Songs 4:10 as charity. Strabo associates the bride's breasts with the "love of God and of neighbor, by which the holy mind nourishes its sentiments, for it is fasten by charity to God, and it pays out what it can to its neighbors."[17] In the twelfth century, Alain of Lille conflates the bride and her breasts with the Virgin Mary as the exemplar of charity:

> The two breasts of the Virgin are the two arms of charity... The two breasts of the Virgin Mary are understood as the two rivulets of charity, by which she loved Christ, one by which she love Him as much as God, the other which she loved Him as much as man.[18]

Similarly, an imitator of Richard of Saint-Laurent in the mid-thirteenth century writes that:

> The breasts of the Virgin can be called love of God and of man, at which she nursed... At her breasts the Son of God has composed the compact of our salvation... The Mother of Mercy nourishes us as well with her double breast... And just like little children we search out these breasts for sucking.[19]

In *The Dialogue with God*, Catherine of Siena endows not only Charity, but also Christ, the Holy Spirit, and the Holy Church with lactating breasts:

> So the soul rests on the breast of Christ crucified who is my love, and so drinks in the milk of virtue... how delightfully glorious *is* this state in which the soul enjoys such union at charity's breast that her mouth is never away from the breast nor the breast without milk.[20]

In the sixteenth century, Saint Teresa compares the soul reborn in Christ to an infant nursing at its mother's breast in *The Way of Perfection*:

> The soul is like an infant still at its mother's breast... It is the Lord's pleasure that, without exercising its thought, the soul... should merely drink the milk which His Majesty puts into its mouth and enjoy its sweetness.[21]

In the visual arts, Charity as a nursing mother makes its first appearance in the thirteenth century. In the process of translating the literary image

---

[17] Walafrid Strabo, *Glossa ordinaria*; quoted in William R. Levin, *The Allegory of Mercy at the Misericordia in Florence: Historiography, Context, Iconography, and the Documentation of Confraternal Charity in the Trecento* (Lapham, MD: University Press of America, 2004), p. 53.

[18] Alain of Lille, *Elucidatio super Cantica Canticorum*; quoted in Levin, p. 54.

[19] Pseudo-Richard of Saint-Laurent, *In Canticum Canticorum Explicatio*; quoted in Levin, p. 54.

[20] Quoted in Yalom, p. 44.

[21] Saint Teresa, *The Complete Works*, trans. and ed. E Allison Peers (London and New York: Sheed and Ward, 1946), vol. 2, pp. 130–1.

into a visual one, the concept of charity gradually becomes more personified and naturalistically rendered. By the fourteenth century, she is depicted as a real woman. Tino da Camaino, for example, sculpts Charity as a robust Italian woman with two very healthy babies nursing at her breasts through small slits in her dress (1321, Museo Bardini, Florence) (see figure 8).

The *Carita Romana* tradition similarly associates Charity with breast-milk, but with some striking differences: rather than a mother nursing her babies, Roman Charity is allegorized as a daughter nursing her father. The allegory derives from the ancient Roman story of Pero who, finding her father, Cimon, incarcerated and condemned to starvation, steals into his prison cell and nurses him back to life with her own breast-milk. Pero's act of filial devotion so impresses the Roman officials that Cimon is released. Pero becomes the model of Roman charity. The *Carita Romana* is a favorite theme in seventeenth- and eighteenth-century painting. Notably, Rubens, whom Charles I commissioned to paint the magnificent allegorical cycle of ceiling paintings for the Banqueting House hall at Whitehall Palace, made several *carita romana* portraits, including *Cimon and Pero* (1630, Rijksmuseum, Amsterdam) (see figure 9).

Fig. 8. Tino da Camaino, *Charity*, 1321.

Carravaggio includes the scene of Pero and Cimon in his painting of *The Seven Great Mercies* (1607, Church of Pio Monte della Misericordia, Naples) (see figure 10).

Under Roman Catholicism, Mary's milk became the object of popular cultic devotion. Placed in vials in churches throughout Europe, so that worshippers could benefit from its miraculous healing powers, Mary's milk was thought to cure blindness, ulcers, and other terrible maladies.[22] In jettisoning the figure of Mary, the Reformation suppressed popular Catholicism's sacramental reverence for Mary's milk. Calvin typifies Protestant skepticism about cultic devotion to the nursing Madonna when, after observing the many samples of the Virgin's milk revered in European churches, he made this wry observation: "Had the breasts of the most Holy Virgin yielded a more copious supply than is given by a cow, or had she continued to nurse during her whole lifetime, she scarcely could have furnished the quantity which is exhibited."[23] The Puritan guidebook authors similarly downplay the sacred mystery of Mary's milk. Unlike the Holy Mother with her miraculous medicinal and intercessory powers, the

Fig. 9. Peter Paul Rubens, *Cimon and Pero*, 1630.

[22] See Dylan Elliot, *Fallen Bodies: Pollution, Sexuality, and Demonology in the Middle Ages* (Philadelphia: University of Pennsylvania Press, 1999), p. 115.
[23] John Calvin, *Tracts and Treatises on the Reformation of the Church*, ed. Henry Beveridge (Grand Rapids, Mich.: W.B. Fredmans, 1958), vol. 1, p. 317.

Fig. 10. Carravagio, *The Seven Great Mercies*, 1607.

new mother in the guidebooks is characterized in moral and scientific terms. The guidebooks' Mary is affiliated with Sarah and Jochobed, among other Hebrew scriptural exemplars of nursing motherhood. For Roman Catholics, Mary's role as spiritual mediatrix between Jesus and the people is unique. For the Puritan guidebook authors, however, the new nursing mother, including Mary, is revered for transmitting physical health and a strong moral character to her offspring—and, more specifically, for constituting the reformed English self in the monist terms celebrated by Hebrew scripture, a point that I explore next.

BIBLICAL MODELS

As just noted, guidebook representations of maternal nurture both reflect and revise received Galenic perceptions of the human body, including the exceptional and uncontainable character of maternal lactation. The Galenic model of human physiology assumes likeness between men and women, especially in relation to sex and reproduction.[24] Both men and women were thought to have seed, or sperm, and breasts, for that matter. Although, in the Galenic schema, the breast was thought to be shared by both sexes, lactation was designated an exclusively female bodily function. In Thomas Vicary's *The Anatomie of the Bodie of Man*, the "generation of milke" categorically distinguishes the female breast from its male counterpart.[25] By distinctly gendering the unisex "pap," maternal lactation disrupts Galenism's mirror-model of human sexuality and reproduction.[26]

If the lactating breast is the exception to Galenism's one sex–one body reproductive paradigm, it forms a foundational site for early modernity's reformation of motherhood—a reformation that domestic guidebooks both inscribe and disseminate. For Puritan guidebook writers like Gouge, Cleaver, and Dod, the categorically female character of lactation not only naturalizes the difference between the two genders, but also reduces "Woman" to her biological role and domestic duties as nursing mother, as the literal and symbolic maternal source of food and nurture, both physical and spiritual, for children, husbands, and the state. In describing maternal breast-feeding as a holy impulse and natural instinct operative in all women, guidebook writers sweep away all socially determined distinctions among women. Placing women, via the lactating breast, on the same plane, the guidebooks' leveling, *tabula rasa* considerations of nursing motherhood divest women of titles, costume, and all other customary external marks of entitlement. The new mother appears in a de-eroticized state of social undress. Stripped of social artifice and ornament, she

---

[24] See Thomas Laqueur, *Making Sex: Body and Gender from the Greeks to Freud* (Cambridge: Harvard University Press, 1990), esp. 25, 37, 38, 171–4. Laqueur's study is contested by, among others, Gail Kern Paster, *The Body Embarrassed: Drama and the Discipline of Shame in Early Modern England* (Ithaca: Cornell University Press, 1993), chapters 2, 4.

[25] Thomas Vicary, *The Anatomie of the Bodie of Man*, 1548. Reissued by the Surgeons of St. Bartholomew's in 1577, ed. Frederick J. Furnivall and Percy Furnivall (London: Early English Texts Society, 1888), p. 55.

[26] As Schwarz aptly puts it in her gloss on Vicary's *Anatomie*: "And even if, as in the Galenic model, women's genitals are imagined to mirror those of men, producing some degree of reproductive mutuality, the maternal breast is an inescapable site of difference." See Katherine Schwarz, "Missing the Breast: Desire, Disease, and the Singular Effect of Amazons," *Body in Parts*, p. 147.

emerges as natural and pure: free from natural and social corruption. Breast-feeding allows the nursing mother to transmit her own purity and well-being to her offspring, thus ensuring not only the physical and moral health of her children, but also the social and spiritual health of the nation.

The influence of Martin Luther can be detected in the guidebooks' frequently reiterated observation that all women, regardless of class, are natural nurturers. In *Lectures on Genesis, Chapters 1–5,* Luther maintains that the female body was designed expressly to nurture children: "To me, it is a source of great pleasure to see that the entire female body was created for the purpose of nurturing children." Nurture, for Luther, is unnatural to men; unlike women, men have absolutely no innate ability to care for their offspring. Whereas "prettily even little girls carry babies on their bosom…Get a man to do the same things and you will say that a camel is dancing, so clumsily will he do the simplest tasks around the baby."[27] Cleaver, Dod, and Gouge also resemble John Streater, Comenius, and other seventeenth-century Reformers by identifying the breeding and education of young children as crucial to the reformation of the nation. For the guidebook writers, however, it is the nursing mother who mediates between the reformed private household and the public realm of the reformed state.

Class, gender, nurture, nature, and national reform thus are among the broad, overlapping concerns that energize guidebook discourse on the purpose and value of maternal breast-feeding. As Robert Pricke puts the question: "To what end doeth the providence of God yeeld unto the woman two Pappes, as it were fountaines, and that in the most comely and fit place of her bodie & besides that, filled them with most sweet and pretious liquor?"[28] Guidebook writers answer this question by invoking the nursing practices of three key biblical matriarchs, Sarah, Jochobed (the Hebrew mother of Moses), and Mary. The guidebooks' most recurrent reference is to Sarah's nursing of Isaac in Genesis 21:7—"And she said, Who would have said unto Abraham, that Sarah should have given children suck? for I have born *him* a child in his old age"—which is read as revealing the undifferentiated nature of maternal breast-feeding[29]. In keeping with their efforts to downplay the socially encoded relations between nursing and status, Cleaver and Dod interpret this passage as providing irrefutable proof that maternal nursing cannot be codified by social

---

[27] Martin Luther, "Lectures on Genesis, Chapters 1–5," *Luther's Works*, vol. 1, ed. Jaroslav Pelikan (St. Louis, Concordia Publishing House, 1958), p. 202.

[28] Robert Pricke, *The Doctrine of Superiority*, section K.

[29] All biblical references are to the Authorized King James Version and are noted in the text.

rank, title, power, and position. Maternal breast-feeding gives expression to and allows (imagined) lateral participation in a national whole, over-riding hierarchical distinctions. Sarah "nursed Isaak, though she were a Princess; and therefore able to have had others to have taken that paines."[30] Sarah reveals the true nobility implied by her name (Sarah is translated from the Hebrew as "princess") by yielding to natural and moral impera-tives for maternal nursing. Moreover, although she has the status and authority to do so, she chooses not to engage others to nurse for her. Gouge glosses the example of Sarah in similar terms:

> *Sarah* gave sucke to Isaak. The example is to be noted especially of the greater sort: as rich mens wives, honourable mens wives, and the like. For *Sarah* was an honourable woman, a princess, a rich man's wife, a beautifull woman, aged and well growne in years, and the mistress of a family.[31]

More explicitly than Cleaver and Dod, Gouge aims to reform mother-hood by instructing women to become new, natural and spiritual rather than traditional, status-conscious interpreters of nursing. He implies that readerly engagement with God's two books, scripture and creation, teaches women to understand breast-feeding as governed by an interlock-ing natural-and-spiritual mandate to nurture—a mandate that obviates the unregulated worldly customs and traditions that hitherto had hierar-chically shaped women's social behavior and identity.

Another Hebrew scriptural passage important to the guidebooks' ef-forts at reforming motherhood is Exodus 2:7–9. These verses narrate that, when the maid of Pharaoh's daughter was sent to find a Hebrew woman to wet-nurse the infant Moses, she was led to Jochobed, Moses' biological mother. Cleaver's and Dod's reading of the text as containing a divine mandate for maternal breast-feeding is illuminating: "So when God chose a nurse for *Moses*, he led the hand-maide of *Pharoas* daughter to his mother: as though God would have none to nurse him but his mother."[32] The wording here refuses to admit an opposition not only between "nurse" and "mother," but also, indirectly, between Pharaoh's royal daughter and Moses' slave-mother. While Cleaver and Dod distinguish the exalted "*Pharoas* daughter" from the lowly Hebrew slave-mother, they also em-phasize that Jochobed, as a natural nursing mother, gains divinely in-spired ascendancy over her royal Egyptian counterpart: "God would have none to nurse him but his mother." By divine mandate, the Egyptian mistress brings the infant Moses to her Hebrew slave such that, unbe-knownst to Pharaoh's daughter, he will be nourished by his biological

---

[30] Cleaver and Dod, *A Godly Form of Household Government*, 1621, sig. P4r.
[31] Gouge, *Of Domesticall Duties*, p. 510.
[32] Cleaver and Dod, 1621, sig. P4r.

mother's milk. For Cleaver and Dod, the example of Jochobed proves that the social distinctions conferred by status and rank are elided by the powerful natural impulses and spiritual mandates that drive all women to provide maternal breast-milk for their babies.

Jochobed's crucial role in nursing the infant Moses has an additional importance to Cleaver and Dod: it ratifies the Hebrew Bible's monist paradigms of the self and the nation. Spiritual, moral, and physical purity are all inseparably one and the same. By supplying her son with natural mother's milk, Jochobed guarantees the spiritual-and-ethnic purity of Moses' native Hebrew identity, despite his Egyptian adoption and upbringing. In part, the guidebooks adopt a Hebrew monist model of the pure self to cleanse the reformed English household of any residual Roman Catholic and other non-native influences. Cleaver's and Dod's glossing of Exodus 2: 7–9 demonstrates that while the maternal drive is an impulse common to women of all nations, classes, and religions, it nevertheless also keeps national boundaries intact. Like Jochobed, the new nursing mother provides a natural safeguard against national cross-breeding, especially when the racial, ethnic, and religious integrity of the nation is threatened or compromised from within or without.

As already noted, Mary provides a third exemplar of nursing motherhood in the guidebook literature. In keeping with Protestantism's efforts to repress idolatrous veneration of the Virgin, guidebook allusions to Mary do not grant her maternal milk the special mystical meaning accorded to it by Roman Catholic tradition.[33] Cleaver and Dod underscore Mary's likeness to rather than her difference from other scriptural mothers: "*Likewise* when the Sonne of God was borne, his Father thought none fit to be his nurse, but the blessed virgin, his mother" (emphasis added).[34] In lowering Mary's exalted maternal status, the presumption of similitude permits Cleaver and Dod to downplay the unique mediating role that the Holy Mother was traditionally thought to play in human redemption. The emphasis on likeness also permits guidebook writers to characterize the blessed Virgin as but one among many in a long line of exemplary nursing mothers. The desire to nurse one's own baby unites of all truly noble women, including Mary—as opposed to wet-nursing, a customary mark of aristocratic status.

---

[33] Frances E. Dolan discusses "the sustained and passionate public debate over [Mary's] status in seventeenth-century England" in "Marian Devotion and Maternal Authority in Seventeenth-Century England," in *Maternal Measures: Figuring Caregiving in the Early Modern Period*, ed. Naomi J. Miller and Naomi Yavneh (Burlington, VT: Ashgate, 2000), pp. 282–92.

[34] Cleaver and Dod, 1621, sig. P4r–v.

Elizabeth Clinton's *The Countess of Lincolnes Nurserie* deploys a similar and equally motivated genealogy of exemplary nursing mothers, but it does so to secure rather than overturn aristocratic measures of status and entitlement:

> Now who shall deny the own mothers suckling of their own children to bee their duty since every godly matrone hath walked in these steps before them: *Eve*, the mother of all the living; *Sarah*, the mother of all the faithful; *Hannah*, so graciously heard of God; *Mary* blessed among women, and called blessed of all ages.[35]

Although almost identical to Cleaver's and Dod's genealogy, Elizabeth Clinton's line of scriptural mothers, from Eve to Mary, promotes maternal breast-feeding over wet-nursing to make an opposing point: that mother's milk, like bloodline, birthright, and other organic measures of aristocratic status, secures traditional forms of nobility and honor.[36] By refusing to nurse their own children, aristocratic women fail to assert their *noblesse oblige*: "to shew the meaner their dutie by our good example." Clinton's book is dedicated "To the Right Honor*able and approved vertuous* La: Briget Countesse of Lincolne," who is set up as a "rare example" for other aristocratic women to emulate: "I wish many [Ladies] may follow you in this good worke." Unlike other women of high status who refuse to breast-feed their own children because they (incorrectly) scorn nursing as rude and vulgar, the Countess has "passed by all excuses, and have ventured upon, & doe goe on with that loving act of a loving mother, in giving the sweete milke of your owne breasts, to your owne child." Clinton hopes all women of "the *Higher* and the *richer* sort" will follow suit, in order to preserve not only their own and their children's virtue and well-being, but also, more generally, that of their class.[37]

Like the guidebooks and *The Countesse of Lincolns Nurserie*, the late sixteenth-century engraving *The power of women* features a genealogy of

---

[35] Elizabeth Clinton, *The Countesse of Lincolne's Nurserie* (Oxford, 1622), sig. B3r. Women Writers Online, Brown University Women Writers Project. For insightful commentary on Clinton's text and on other mothers' advice books, see Valerie Wayne, "Advice for Women from Mothers and Patriarchs," in *Women and Literature in Britain, 1500–1700*, ed. Helen Wilcox (Cambridge: Cambridge University Press, 1996), pp. 56–79.

[36] For an alternative reading of Clinton on class and nursing, see Margaret Thickstun, "Milton Among Puritan Women: Affiliative Spirituality and the Conclusion of 'Paradise Lost,'" *Religion and Literature*, 36.2 (Summer 2004): 1–23. For Thickstun, Clinton's advocacy of maternal breast-feeding is "a potentially subversive activity," which challenged the fundamental interests of her class: "Because during this period breastfeeding was understood to have contraceptive effects, aristocratic families who desired many heirs prevented their women from nursing their own infants" (12).

[37] *The Countess of Lincolnes Nurserie*, sig. C2r; sig. A2r.

scriptural mothers, but of a very different kind.[38] Whereas, for opposing reasons, the guidebook authors and Elizabeth Clinton emphasize the continuity between the Old and New Testaments, *The power of women* deploys pictorial typology to emphasize the opposition between the two dispensations: the sterility of the old era against the fertility of the new one. Traditionally, the passage from the old to the new dispensation is allegorized through birth imagery. The Virgin, as the mother of Jesus and all those who are reborn in the Son, mediates between the two eras, setting the barrenness of the Old Testament against the fruitfulness of the New Testament.[39] *The power of women* sets up the traditional barren–fruitful, Old–New Testament dichotomies. But, rather than associate the passage from the Hebrew to the Christian era with spiritual childbirth and the Holy Mother, it allegorically associates this epochal shift with the new sacred–secular nursing mother.

The new nursing mother is the central focus of *The power of women*. She opposes the ungodly, childless, and non-maternal women featured in two Old Testament events: the Solomon episode in 1 Kings 11: 1–10 and the shearing of Samson's locks in Judges 16:19. The Solomon passage, which tells of how Solomon's seven hundred wives "turned his heart after other gods and persuaded him to worship idols," is represented in the right background of the engraving. The idol is shown to be a treacherous but alluring Oriental woman—a naked goddess, an Isis perhaps—whose outstretched arm holds a scepter and a crown, negatively paralleling the right arm of the nursing mother in the foreground. The left background pictures the Philistine Delilah, a phallic knife in hand, cradling Samson as she shears him of his locks and great strength, at once emasculating and infantilizing him. The engraving turns spousal betrayal into a perversion of maternal succor. These two images forge links between various sorts of female lack and perversity: vicious mothering, feminine seduction, failed Hebraic manhood (coded female), and orientalized female idolatry. They are corrected and effaced, driven to the literal margins of the engraving's construction of scriptural history, by the central figure of the nursing mother. At her feet lie broken shields and swords, a severed crown, and a spilled money coffer, emblems of temporal power overcome by the spiritual strength transmitted by her breast-milk to her (male) child. Mothers'

---

[38] Françoise Borin in "Judging By Images," in *A History of Women in the West*, ed. Natalie Zemon Davis and Arlette Farge, vol. 3 (Cambridge, MA: Belknap-Harvard University Press, 1993), p. 209, and Madlyn Kahr in "Delilah," *The Art Bulletin* 54.3 (1972), 297–8 attribute this work to Marten De Vos and identify its title as "Allegory of the Power of Women."

[39] I rely here on Cristelle L. Baskins, "Typology, Sexuality, and the Renaissance Esther," in *Sexuality and Gender in Early Modern Europe: Institutions, Texts, Images*, ed. James Grantham Turner (Cambridge: Cambridge University Press, 1993), pp. 31–54.

milk is a categorically female source of redemptive power; it provides a divinely mandated form of nourishment for the infant kings who will spiritually overcome the all-too-fleshy Solomons and Samsons of the old corrupt dispensation and lead humankind into a reformed new world, a world that is also a nursing motherland as a latter-day promised land of milk (and honey) proclaimed by biblical prophecy.

In its depiction of Solomon, *The power of women* reflects the Reformation's complex re-assessment of the Song of Songs from Hebrew scripture. Solomon's Song inspired at least two very different Protestant interpretations. On the one hand, the Song's allusions to Solomon's legendary riches, especially those amassed through trade with the East, made the Hebrew king "an exemplar of the sage colonial ruler."[40] On the other hand, the Song of Song's celebration of the power of passionate erotic attachment provided a legitimating framework for competing early modern ideas about the relations between eros and marriage, and between cupiditas and caritas.[41] Phillis Trible suggests that the Song's recurrent references to the mother's body, to the mother/child pair without "a single mention of the father, underscore the prominence of the female in the lyrics of love."[42] Protestant exegesis translates the Song's erotic celebration of the maternal and the feminine into a spiritual allegory of the church, scripture, and the ecstatic rapture of the true believer. As David Leverenz observes, for Puritans, "the Bible was God's milk, the minister was the breast at which the congregation suckled," the Song of Songs legitimated this allegory.[43] "It is said, (Cant. 8.8), *We have a little Sister that has no Breast, what shall we do for her?* So there are little places (and some considerable ones too) in *New-England*, that have no Breasts, no Ministers from whom they may receive the sincere milk of the Word," writes Increase Mather in 1697.[44]

[40] Kim F. Hall, *Things of Darkness*, pp. 108, 110. Hall reads Solomon's poem, the Canticles, as providing a scriptural prototype for the English sonnet's negotiations of gender and race: "Solomon's poem provides an obvious model in the poetry of praise for the subordination of the female to male power, the meeting and conversion of foreign female difference to Christian, male standards."

[41] Krier, *Birth Passages*, p. 88. "Far from embracing merchant trading," Spenser in the blazon of "Ye tradefull Merchants" "takes pains to distinguish his speaker's ferocious willingness to trade from the Song's reveling in exotic materials brought by merchants." For Krier, Spenser's response to the Song in his sonnets highlights his desire to "discover or create a language sufficient to the passionate, hymeneal love he hopes to establish with the lady. To establish the poet's right relation to language would be simultaneously to establish the lovers' right relation to each other."

[42] Phillis Trible, *God and the Rhetoric of Sexuality* (Minneapolis: Fortress, 1978), p. 158.

[43] Leverenz, p. 1.

[44] Increase Mather, *David Serving His Generation* (n.p., 1697), p. 31.

The same allegorical reading of the Song also informs Puritan sermons and commentaries on other scriptural passages. For instance, for Thomas Hooker, Isaiah 66:11 links the relationship between gospel and the church to the intimate bond between the nursing child and its mother:

> The Church is compared to a childe, and the brests are the promises of the Gospell; now the elect must suck out and be satisfied with it, and milke it out:...Ah beloved, this is our misery, we suffer abundance of milke to be in the promise, and we are like wainly children, that lye at the brest, and will neither sucke nor be quiet.[45]

As Leverenz observes, "Drinking the word became a metaphor so pervasive in Puritan culture that Cotton entitles his standard catechism for New England children: *Spiritual Milk for Boston Babes In either England. Drawn out of the Breasts of both Testaments for their souls nourishment.*"[46] Like Cotton, William Dickinson (1628), Hugh Peters (1641), and James Naylor (1661) all wrote religious tracts entitled *Milk for Babies.*[47]

Like the New England ministers, *The power of women* relies upon the pervasive metaphor between maternal nursing and scriptural exegesis. But, it also is especially notable for its emphasis on the tenderness with which the new mother executes the harsh Pauline doctrine of supersession. Drinking in the Word renders Hebrew scripture and Hebrew manhood, which is shown to encompass both Samson's strength and Solomon's wisdom, non-nourishing and insubstantial. The nursing mother and child dominate the field of vision, marginalizing the Hebrew scriptural scenes, both of which highlight the decadence and failure of Hebraic patriarchy and the Law. The triumph of Christian charity—allegorized by the nursing mother—over Judaic law is additionally expressed by the ways that the nursing mother of the new era supersedes the evil temptresses of the old era. Whereas Solomon's foreign wives and the Philistine Delilah overpower the sage rulers and judges of the Old Testament by feeding their carnal desires, the new nurturing mother serves her child by nurturing him from the breast with the milk of the gospels. The new Christian mother replaces Solomon; her new nurture domesticates the unruly appetites and oriental excess of the old world, for which, as just noted, Solomon serves as an exemplar. By drinking in the Word by suckling at his mother's breast, the nursing infant surpasses King Solomon in true wisdom.

The nursing mother (or Christian charity) additionally towers over her foreign female counterparts. These tiny figures are consigned to the

---

[45] Thomas Hooker, *The Soules Vocation* (London, 1638), p. 306.
[46] Leverenz, p. 2.
[47] See Boyd Berry, "The First English Pediatricians," p. 575.

drawing's Oriental peripheries, while she occupies the new domestic, Occidental center. In *The power of women*, the nursing mother is a gateway figure. A secular substitute for the Virgin Mother, she mediates between the unregenerate past and salvific future, between the old Israel and the new Israel, between the dark corners of the earth and the enlightened Christian community the world over. The bright collective future belongs to new nursing mother's breast-fed offspring: those nourished by the milk of the gospels. I shall have more to say about *The power of women* and its pictorial typology in relation to Milton's gendered view of the typological similarities and differences between the Old and New Testaments in Chapter 4.

DISENCHANTING MATERNITY, THE NEW SCIENCE, AND THE RISE OF THE "NATURAL" MOTHER

In *The power of women*, pictorial typology unites the secular and sacred dimensions of the nursing mother. In the guidebooks, biblical exegesis and the new science constitute the new mother's role as a natural instrument for forging the moral character of the reformed nation and national subject. Natural science complements the guidebook authors' exegetical justifications of maternal breast-feeding. As already noted, the guidebooks' draw extensively on Galenic theories of the four humors; at the same time, however, they also appeal to experience and to the truth-claims of the new philosophy when they define breast-feeding as the function of inductively derived natural laws:

> We see by experience that every beast and every fowle is nourished and bred of the same that beare it: onely some women love to be mothers, but not nurse. As therefore every tree doth cherish and nourish that which it bringeth forth: even so also it becometh naturall mothers to nourish their children with their owne milke.[48]

By describing maternal breast-feeding as a natural law governing all women, guidebook writers like Cleaver and Dod help to define breast-feeding as a manifestation of both motherly affect ("incredible love") and biological drive. This definition roots maternal nurture in both maternal nature and Mother Nature: nurture *is* nature. Reconceived as a natural phenomenon, maternal nurture could be clearly differentiated from wet-nursing and other vulgar customs and traditions, and newly illuminated and normalized under the dry light of Baconian reason. The natural law

---

[48] Robert Cleaver and John Dod, *A Godly Form of Household Government*, 1621, sig. p4r.

of nurture separates the reformed nursing mother from the unrecon-
structed one, who is associated with the unnatural chaos and violence of
pre-rationalized experience. As Guillemeau maintains in *The Nursing of
Children*, there is "no difference between a woman that refuses to nurse
her own childe, and one that kils her child."

Similar divisions of the good mother from bad mother, along the axis
of the natural and the unnatural, structure early modern obstetrical discourse. The rise of early modern obstetrics roughly parallels the printing
history of the domestic guidebooks. William Harvey's work on obstetrics,
De Partu, was not published until 1653, although his first anatomy lec-
ture on female reproduction dates to April 16, 1616, roughly concurrent
with the publication of Gouge's *Of Domesticall Duties*, which went
through twelve printings over the course of the seventeenth century, ten
after 1626. Ambroise Paré's important work on obstetrics, which appeared
in the original French in 1549, was not translated into English until 1634.
But Guillemeau, Paré's pupil, disseminated his teacher's ideas in his 1612
volume, *Childbirth, or the happy deliverie of women*, to which the aforementioned *The Nursing of Children* was affixed.[49]

Early modern obstetrics aimed to advance knowledge about the birth
process by translating superstitions, folklore, and other popular myths
about motherhood into empirical truths. But this translation inevitably
failed to render conception completely accessible or analyzable in scientific terms. Lacking the technology needed to observe conception, which
is visible only at the microscopic level, early modern obstetrical investiga-
tors could not empirically pinpoint when or how life began. The biological moment of life's beginning remained shrouded in mystery and closed
off to the new science. For physicians and surgeons like the eminent William Harvey, the murky, inaccessible, and seemingly invisible reproductive matrix proved to be what Richard Wilson describes as "the blind
spot" of medical investigation.[50] As Wilson observes, Harvey attributed

---

[49] For an erudite history of early modern women's medicine in England and France, see
Helen King, *Midwifery, Obstretrics, and the Rise of Gynaecology: The Uses of a Sixteenth-
Century Compendium* (Burlington, VT: Ashgate, 2007).

[50] Richard Wilson discusses Harvey's obstetrical experiments in "Observations on English Bodies: Licensing Maternity in Shakespeare's Late Plays," in *Enclosure Acts: Sexuality,
Property, and Culture in Early Modern England*, ed. Richard Burt and John Michael Archer
(Ithaca: Cornell University Press, 1994), pp. 121–50. Unlike Harvey's obstetrics, which
cohere with his monarchist view of the body politic, the celebrated doctor's discovery of the
circulation of the blood, according to Christopher Hill and John Rogers, reflects commonwealth views of the ideal political body. See Christopher Hill, "William Harvey and the
Idea of Monarchy," *Past and Present*, 27 (1964); reprinted in *The Intellectual Revolution of
the Seventeenth Century*, ed. Charles Webster (London: Routledge, 1974), pp. 160–81, and
John Rogers, *The Matter of Revolution: Science, Poetry, and Politics in the Age of Milton*
(Ithaca: Cornell University Press, 1996), pp. 16–38.

his failure to detect the biological beginning of life to the cultural and legal limits imposed on anatomists. He despairs in *Anatomical Exercitations, Concerning the Generation of Living Creatures* that "there is more difficulty in the search into the *Generation* of *Viviparous Animals*: for we are almost quite debarred of dissecting the humane *Uterus*."[51] Percivall Willoughby similarly laments that the advancement of obstetrical knowledge in vulgar England could not keep pace with more enlightened countries such as Holland, where "they have privileges we cannot obtain. They open dead bodies without the mutterings of their friends. Should one of us desire such a thing, an odium of inhumane cruelty would be upon us from the vulgar." The customary banishment of men from the exclusively female birth room, over which midwives and nurses presided, further limited what the Harveian observer could see. Willoughby writes how in order to witness a birth he "crept into the chamber on my hands and knees, and returned, so that I was not perceived by the lady."[52]

Banished from the birth room and prevented from anatomizing human wombs, Harvey applied himself to cracking open chicken eggs and dissecting the wombs of pregnant deer to see if he could discern life's origins. He chose chicken eggs because they were so plentiful and inexpensive, and deer because, at the bequest of his patron, Charles I, a dedicated hunter and amateur scientist, he was given an ample supply of dead bodies to dissect:

> Our late Sovereign King *Charles,* no sooner as he became a Man, was want for Recreation, and Health sake to *hunt* almost every week. I had a daily opportunity of dissecting the [deer] by the favour and bounty of my Royal Master (whose Physitian I was), and who was himself much delighted in this kind of curiosity, being many times pleased to be an eye witness.[53].

Still, even with the patronage of his monarch, Harvey could not bring the secrets of the womb to light. The female reproductive matrix remained a dark, murky mystery. In *Anatomical Exercitations, Concerning the Generation of Living Creatures*, Harvey writes of conception: "wee meet with more things wanting out wonder.... It is indeed a dark, obscure business...dark matters." "I have still thought much more remained behind, hidden by the dusky night of nature, uninterrogated." Harvey

---

[51] William Harvey, *Anatomical Excercitations, Concerning the Generation of Living Creatures: To Which are Added Particular Discourses, of Births, and of Conceptions, &c.* (London, 1653), p. 391.

[52] Percivall Willoughby, *Observations in Midwifery*, quoted in Wilson, pp. 125, 127.

[53] William Harvey, *Anatomical Exercitations*, p. 396. This collaboration between king and physician suggests the new confluence between medical knowledge and political power that cultural historians such as Foucault date to the early modern period.[53]Michel Foucault, *The Birth of the Clinic: Archaeology of Medical Perception*, trans. Alan Sheridan (London: Tavistock, 1976).

optimistically embarks on his medical adventures, hoping to discern what man "was in his Mothers Womb, before he was this *Embryo*, or *Foetus*; whether *three bubbles?* or some *rude* and *indigested lump?* or a *conception*, or *coagulation* of *mixed seen?* or whether any thing else?" But he receives no answers to his question, encountering only deeper mysteries: "For there is a far greater, and diviner mystery in Generation, then a *bare* assembling, altering, and compounding of *Parts.*" The great scientist's dissections allow for an accounting and assembling of parts, but the parts do not add up to the whole; "the *Whole*," Harvey concludes, "is made and discovered *before* its *parts*" (emphasis added). It is precisely to this "before" that Harvey, to his terrible disappointment, fails to gain access. This anterior temporal space eludes his grasp; it is both "there" and not "there": "I say *there* is no *Sensible thing* to be *found* in the *Uterus,* after coition; and yet there is a necessity, that something should be *there*, which may render the female fruitful." The new science's progressive vision of expanding knowledge in an intellectual future with no set horizons is brought up short in the face of the impenetrable maternal darkness of that which came before.[54]

In the end, Harvey claims victory by converting failure into triumph. He concludes that the missing "something [that] should be there" but could not be found in the uterus after coition is in fact there after all. But rather than a "*Sensible thing*," this "something" is no thing at all, but rather nothing less than the invisible imprint "of the *Eternall minde* of the *Divine Creator* which is imprinted in Things, [and] doth create the *Image* of itself in Humane Conceptions." What seems to be an absence is in fact the invisible trace of the divine presence. Hence, rather than leave the anatomist with specific knowledge of parts, but no insight into the whole, dissection leads directly to Truth, or at least to the shadow or "certain adumbration" of Truth, itself: "not only the knowledge of those less considerable secrets of Nature, but even a certain adumbration of that Supreme Essence, the Creator." Harvey concludes that while conception takes place in the body, the beginning of life is itself an immaterial phenomenon, akin to thought: "The *Conception* of the Egge, or the *Uterus* is (in some sort) like the Conception of the Brain it selfe."[55]

In the guidebooks, the unfulfilled Harveian fantasy of bringing the maternal body and the reproductive matrix to light is played out more successfully, but in a different, political key. In these texts, the obscure, devalued space of the nurturing, maternal-centered, private household is transported to the center of a new cultural map, where the new moral and

---

[54] Harvey, *Anatomical Exercitations*, pp. 539, The Preface, 230.
[55] Harvey, *Anatomical Exercitations*, pp. 547, 550, The Epistle Dedicatory, 554, 547.

medical value of maternal nursing could be applied most broadly. Thus, while Harvey focuses his attention on the womb, the guidebook authors concentrate on the breast. Unlike childbirth, nursing was understood to be a matter of deliberate choice—an assumption that coheres with Puritan emphasis on the role of conscience in determining moral character. As Cleaver and Dod emphasize, Sarah "nursed Isaak, though she were a Princess; and therefore able to have had others to have taken that paines."[56] The notion that nursing produces a second birth, which eradicates the child's biological likeness to its parents, also influences Puritan emphasis on the breast over the womb. (It is important to remember, however, that the substitution of nursing for childbirth as key to the offspring's identity does not always break down over class or sectarian lines. Arguing that aristocratic women should assert their own agency rather than obey custom, Elizabeth Clinton similarly emphasizes the element of choice in maternal nursing, although for different reasons, as we have already seen.)[57]

In the guidebooks, mothers who choose to attend to the natural and divine call to nurse exhibit their social value and utility in two key ways: they fix class parameters, and they protect against the threat of foreign mixture and cross-breeding. As suggested in my Introduction, guidebook writers like Gouge worried about the ill-effects of wet-nursing because they were convinced that breast-milk forged suspect but ineffaceable bonds between wet nurses and children: "such children as have sucked their mothers breasts, love their mothers best: yea we observe many who have sucked others milke, to love those nurses all the daies of their life."[58] This perception of nurse-milk's competition with and perversion of parentally transmitted identity is familiar to us from Shakespeare's plays, for example, *The Winter's Tale*.[59] Leontes demands that Mamillius (whose name emphasizes his attachment to and determination by the maternal breast) be removed from the powerful shaping influence of what he perceives as Hermione's treacherous, unlawful body. He laments that "[though] he does bear some signs of me," his son has too much of her adulterous, female, and so unstable and illegitimate "blood in him." His only solace is: "I am glad that you did not nurse him" (2.1, 57–8, 56). The

---

[56] Cleaver and Dod, *A Godly Form of Household Government*, 1621, sig. P4r.

[57] Jacqueline T. Miller, "Mother Tongues": 192.

[58] Gouge, *Of Domesticall Duties*, p. 512.

[59] See Gail Kern Paster's definitive chapter on "Quarreling with the Dug, Or I am Glad You Did Not Nurse Him," in *The Body Embarrassed: Drama and the Discipline of Shame in Early Modern England* (Ithaca: Cornell University Press, 1993). Paster describes the later events of *The Winter's Tale* as "the masterplotted representation of the generic desires and intrapsychic traumas of wet-nursed children," p. 274. On the association of nursing with witchcraft in the play, see Kirstie Gulick Rosenfild, "Nursing Nothing: Witchcraft and Female Sexuality in *The Winter's Tale*," *Mosaic*, 35 (2002): 95–112.

guidebooks similarly suggest that children put out to wet nurses, or breast-fed by unfit mothers, not only are more likely to die than those offered tender, maternal care and nourishment—"The number of nurse children that die every yeare is very great"—but these children are also apt to experience a kind of social death, the implications of which, as we have seen, Elizabeth Clinton addresses in *The Countesse of Lincolns Nurserie*. Unlike maternal breast-feeding, which keeps the flow of milk within the same family and class, wet-nursing creates milk-lines that threatened to mix families and blur or subvert the borders separating the upper-class infant from the lower-class nurse.

Such attempts at providing more efficient plumbing for maternal milk-flow are oddly contiguous with contemporaneous debates about London's water supply and the various private and civic strategies devised to make the flow of water more efficient and less susceptible to leakage. Indeed, discourse on English plumbing has quite a direct bearing on the nursing debates and the guidebooks' preoccupation with directing and enclosing mothers' milk within traditional class and national parameters, since, as Jonathan Gil Harris points out, the plumbing techniques suggested and even practiced by city officials and private landowners like Samuel Rolle and Edward Forset for damning up or circulating water to create optimal usage "were very much mediated by patriarchal constructions of gender and parenthood," which coded the civic water supply as unruly, feminine flow requiring the control and containment that male pipes and castellated conduits could exert.[60] Like the reformed maternal breasts in the guidebooks, the public fountains, as Rolle puts it, "ministereth…nourishment to every part";[61] the flow of water, like the flow of breast-milk, required direction, regulation, and enclosure to protect against leakage and pollution.

This implied analogy between the reformed maternal breast and the public fountain is made explicit in James Harrington's *Oceana*:

> if a river have had many natural beds or channels, to which she hath forgotten to reach her breast and whose mouth are dried up or obstructed, these are the dams which the agrarian doth not make but remove; and what parched fortunes can hereby hope to be watered by there only, whose veins have drunk the same blood have a right in nature to drink the same milk.[62]

[60] Jonathan Gil Harris, "This is Not a Pipe: Water Supply, Incontinent Sources, and the Leaky Body Politic," in *Enclosure Acts*, p. 206.

[61] Quoted in Harris, "This is Not A Pipe," p. 205.

[62] James Harrington, *Oceana*, in *The Political Works of James Harrington*, ed. J.G.A. Pocock (Cambridge: Cambridge University Press, 1977), p. 468. I discuss this passage in a different context in "Female Preachers and Male Wives: Gender and Authority in Civil War England," in *Pamphlet Wars: Prose in the English Revolution*, ed. James Holstun (London: Frank Cass, 1992), pp. 122–3.

In Harrington's formulation, the state itself and its public works appropriate and improve upon the natural flow of water/maternal milk to achieve optimal social circulation among the English citizenry. Harrington insists that national sanguinity and identity depend on unblocked access to a carefully regulated and well-managed, national water/breast-milk supply. Those of "the same blood" have a natural right to "the same milk" and should not be forced to drink from a fount that is different, strange, or polluted.

In their attempts to rearticulate the maternal, the domestic guidebooks and affiliated texts illuminate the very specific socio-political connotations that the natural begins to acquire in the early modern period. As Stephen Greenblatt observes, over the course of the sixteenth and seventeenth centuries the natural replaces the sacred as the conceptual category governing early modern English attempts to map the boundaries between England and the radically different cultures it increasingly encountered through stepped-up trade, missionary work, and early acts of colonization. As Greenblatt notes:

> the sixteenth and seventeenth centuries also saw the beginning of a gradual shift away from the axis of sacred and demonic and toward the axis of natural and unnatural... but the natural is not to be found, or at least not reliably found, among primitive or uncivilized peoples.... The stage is set for the self-congratulatory conclusion that European culture, and English culture in particular, is at once the most civilized and the most natural.[63]

Greenblatt does not extend his analysis of the shift from the "sacred and demonic" to the "natural and unnatural" into the domestic sphere, but the same conceptual transformations can be located there. The guidebook writers make a special point of describing the nursing mother as "naturall"—in at least partial recognition of the new meaning and value of this term as a synonym for civilized, English culture. In guidebook formulations, the natural nursing mother shores up the self-congratulatory construction of English identity by interpellating her suckling child into the natural order and thus, in the same gesture, into the civilized order of English culture. In contrast, the unnatural mother (the wet nurse or the mother who refuses to nurse her own babies) renders children both unnatural and uncivilized or, in short, not English. By nursing her children, the new natural mother seamlessly translates nature into culture and birth identity into national identity. Through the same power of translation (and translation of power) she also bears the potential of transmuting the old nation's bonds of blood and kinship into the new nation's abstract

---

[63] Stephen Greenblatt, "Mutilation as Meaning," in *Body in Parts*, pp. 230, 236.

and anonymous forms of affective social attachment. Through the power of maternal breast-feeding to naturalize and, hence, civilize children/subjects, the new, natural mother perhaps most fully reveals her social utility to the reformed nation.

Although the natural nursing mother helps to transform the old nation into the new one, she nevertheless also consolidates traditional constructions of English identity. The politically conservative vision of the natural mother's role gains strength from the Galenic conception of breast-milk as white blood—"nothing else but blood whitened," as Guillemeau maintains. Politics and religion, however, can be irrelevant when considering the broader currency of Galenism. Just as aristocrats like Elizabeth Clinton adhere to the humoral theory that mother's milk is white blood, so too do the more progressive Puritan guidebook authors: "Experience teacheth, that God converteth the mothers bloud into the milke where with the child is nurse," write Cleaver and Dod. With its potential to mix bloodlines, nurse-milk carries the threat of moral and proto-racial contamination and contagion. "Now if the nurse be of an evill complexion," write Cleaver and Dod, "… the child sucking of her breast must needs take part with her."[64] By "evill complexion," Cleaver and Dod refer to moral character, as in the humoral-based theory of the four temperaments. Galen's *Pericraison; De temperamentis, On the Temperaments* (the standard authority until the sixteenth century), made it possible to understand personality and character in bodily terms as mixtures or temperaments (*crasis*) according to the predominance of one of the humors in the body. Just as there were four humors, there also were four temperaments. Despite the powerful influence of Galen, however, it is also possible to detect the new scientific or empirically based connotations that "complexion" begins to acquire in the period, when the term alludes for the first time to hair and skin color.[65] As deployed in the passage from Cleaver and Dod, "evill complexion" suggests an incipient slide from a humoral notion of temperament to a proto-racialized notion of identity that paints the skin-color of the bad nurse in dark hues.

More generally, anxieties about transgressive blood-milk mixtures reflect concerns about class, birthright, bloodline, and the transmission of status, title, and property. It is with these concerns in mind that Queen Anne, the wife of James I, lines up with the advocates of maternal breast-feeding: "Will I let my child, the child of a king, suck the milk of a subject and

---

[64] Cleaver and Dod, 1621, sig. P4v.
[65] See "complexion" (entry 4), *Oxford English Dictionary*, which dates the term's first reference to "the natural colour, texture, and appearance of the skin, *esp.* the face," to 1568.

mingle the royal blood with the blood of a servant?"[66] Outside the English context, the slippage between class and proto-racialized anxieties about wet-nursing can be much more explicit. For example, Spanish legislation prohibited "a New Christian woman to serve as a wet nurse for the royal children because her milk is polluted since she is of the despised and accursed race [of Jews]."[67] However, as noted, the very same pure milk–pure blood equation allows the maternal breast to serve as a metaphor for a natural-rights utopia such as Harrington's *Oceana*: "those and what parched fortunes can hereby hope to be watered by there only, whose veins have drunk the same blood have a right in nature to drink the same milk."[68]

In all of the aforementioned examples, whether conservative or reformed, the need to naturalize maternal breast-milk underscores just how difficult it had become to preserve blood as a stable marker of social status and political power. The contested significance of blood as a governing determinant of personal and political self-worth makes the discursive drive to legitimate maternal nurture all the more urgent for writers of all political vantage points. For republicans like Harrington and Milton, the subversive, second creation made possible by maternal breast-feeding helps to make merit, as opposed to bloodline, the chief measure of social position and communal belonging. In contrast, as we shall see in the next two chapters, James I, Charles I, and Shakespeare in *Macbeth* draw out the conservative implications of natural nursing motherhood. To strengthen traditional bloodline and birthright justifications of power and position, they refashion the king in the venerable image of the Old Testament's nursing father in Isaiah. In turn, Reformers appropriate Stuart images of the nursing father and refashion them into anti-monarchical emblems of male political power.

LANGUAGE AND LACTATION

Language represents another equally crucial arena in which idealized representations of nursing mothers mix with competing visions of the national future. In a fascinating essay, Linda Phyllis Austern comments on

---

[66] Quoted in Marilyn Yalom, *A History of the Breast* (New York: Knopf, 1997), p. 85. In her reading of *Macbeth*, Deborah Willis argues that "it is useful to distinguish between 'mothers'...whereas Lady Macbeth...embodies aspects of the aristocratic mother [for whom breast-feeding represented an upper-class stigma], the witches more closely parallel the lower-class nurse," *Malevolent Nurture*, p. 217. But, as the examples of Queen Anne and Elizabeth Clinton suggest, these two different categories of mothers are complicated when aristocratic women begin to argue that they should nurse their own babies as a way to prevent their offspring's upper-class blood from mixing with the wet nurse's lower-class milk.

[67] Quoted in James Shapiro, *Shakespeare and the Jews* (New York: Columbia University Press, 1996).

[68] Harrington, p.468.

the close relationship between the maternal breast and the nurturing of sound through music. "Musica," she notes, "had been embodied as a woman since antiquity," but in the Renaissance, she was specifically figured as a nursing mother. Cesare Ripa's *Iconologia* "presents the most important representation of lactation in service of music." His "emblem of Poesia reinforces the inseparability of music and poetry" by presenting the art as "le mammelle piene di latti"—with her lush left breast bared. As Austern documents, "this emblem exerted an inestimable influence on allegories of musical portraiture for over a century and a half."[69] Music and sound itself were imbibed at the mother's breast. This perception strengthened the commonplace linkage of breast-milk to language, as in the mother tongue. In an early seventeenth-century tract, William Austin writes that language originates from the nursing mother–child relation. Speech recapitulates the infantile pleasure of imbibing mother's milk: "For in our infancy, we learne our language from them. Which men (therein not ingratefull) have justly termed our *Mother tongue*."[70] The same intimacy between language and nursing is corroborated rather differently by Montaigne in "Of the Education of Children." He describes how he learned Latin as if from the breast: "while I was nursing and before the loosening of my tongue."[71] Educated from infancy by his father and a male tutor, Montaigne drinks in Latin at the same time as he drinks in his nurse-milk. Nursing thus allows the infant essayist to gain fluency in his father tongue (Latin) as well as in his mother tongue, without his ever confusing the two.

More generally, seventeenth-century discourse on breast-feeding is related to the period's highly motivated battles over the very nature of language itself. As Sharon Achinstein observes, "The contests over political and social authority were represented by figures for clashes in language because public expression was becoming the means by which political and social differences were made known."[72] Schemes for a universal language that, such as those proposed by John Webster, would "have repaired the ruines of *Babell*", proliferated throughout the seventeenth century.[73] The linguistic reform movement found its initial source of inspiration in Francis Bacon's desire to create a system of "Characters Real, which express

---

[69] Linda Austern, "'My Mother Musicke': Music and Early Modern Fantasies of Embodiment," in *Maternal Measures*, ed. Naomi J. Miller and Naomi Yavneh, pp. 241, 258.

[70] Quoted in Austern, p. 244.

[71] Michel de Montaigne, "Of the Education of Children," in *The Complete Works of Montaigne,* ed. and trans, Donald M. Frame (Palo Alto: Stanford University Press, 1956), p. 128.

[72] Sharon Achinstein, "The Politics of Babel in the English Revolution," in *Pamphlet Wars: Prose in the English Revolution*, ed. James Holstun, p. 17.

[73] John Webster, *Academiarum Examen* (London, 1654), p. 25.

neither letters nor words in gross, but Things or Notions."[74] The new world order to be ushered in by the new science required a new universal language. For Bacon, this new language would not only aid in international communication, but it would also help to unite the intellectual and political spheres, thought and action, into one all-inclusive world system and a wholly integrated mode of being. A universal language would help humankind to recover the lost *integritas* of its Edenic estate. English language reformers such as John Webb tried to realize Bacon's vision of regaining Paradise by reviving "the language spoke by our first Parents...before the *Confusion of Tongues* at *Babel.*" For Webb, this original language was the "*Lingua Humana, the Humane Tongue.*"[75] At first glance, the expansive national and international concerns of language reformers like Bacon, Webb, and Webster appear distant from the circumscribed, domestic interests of guidebook writers and other authors interested in reforming the English home and mother. Despite their conspicuous differences, however, both reformist projects share the same politically motivated desire to gain access to primal forms of communication, to the lost origins of language itself. What, for Webb, is the "*Humane Tongue*" is the "*Mother tongue*" for Austin. As Austern observes, the first sound to echo in the infant's ear, the mother's tender voice was thought to linger forever, "an auditory caress of sublime, all-encompassing love and the remembered taste of true nourishment."[76] As the guidebooks proclaim time and again, wet-nursing binds children to an unnatural breast and so also to a non-native maternal voice and mother tongue, confusing their natural allegiances to family and nation. It is this very intimate and almost primal "*Confusion of Tongues*" that the guidebook writers try to redress.

Like William Austin, Spenser identifies maternal breast-feeding as the key to language acquisition. For him, language, culture, and national identity are inseparably related—and all emanate from the breast. In *A View of the Present State of Ireland*, he offers perhaps the clearest early modern representation of the cross-currents between the new ideology of maternal nursing and England's emergent linguistic nationalism.[77] For

---

[74] Francis Bacon, *The Works of Francis Bacon*, ed. James Spedding, Robert Leslie Ellis, and Douglas Denon Heath, 14 vols. (London: Longman, 1860), pp. iii, 399–400.

[75] John Webb, *An Historical Essay* (London, 1669), pp. 16–17.

[76] Austern, p. 243.

[77] Spenser's discussion of Irish wet nurses has been discussed in several recent studies. In *The Reformation of the Subject: Spenser, Milton, and the English Protestant Epic* (Cambridge: Cambridge University Press, 1995), p. 12, Linda Gregerson argues that in Spenser's *A View of the Present State of Ireland*, the nursing breast functions as "a figurative gauge for the theory and practice of colonial rule." In *Things of Darkness*, p. 146, Kim Hall reads this passage as presenting "English fears of alternative social structures."

Spenser, the Irish wet nurse creates unwholesome and unnatural linguistic and cultural mixtures, which fatally blur the borders between English and Irish identity. Linda Gregerson observes that, in keeping with "the logic of sixteenth- and seventeenth-century English colonialism," Spenser treats the relations between language and lactation in the context of "apt and recalcitrant learners."[78] For Spenser, Irish wet-nursing results in a serious impairment of both English learning and the learning of English, collapsing the divide between the colonizing English and the colonized Irish:

> for first the childe that sucketh the milke of the [Irish] nurse must of necessity learn his first speech of her, the which being the first that is enured to his tongue is ever after most pleasing unto him, insomuch as though he afterwards be taught English, yet the smack of the first will always abide with him, and not only of the speech, but of the manners and conditions.[79]

Just as English landowners who, by adopting Irish customs and assuming Irish names, "bite off her dug from which they sucked life," their children, given over to Irish wet nurses, prefer the delicious "smack" of the barbarous Irish they suck in at the breast to the proper English, which they are later taught with considerable difficulty. The children miss not only the delicious taste of the Irish language that they imbibed as babes at the breast of their Irish nurses, but also the love and tenderness they drank in with their breast-milk: "the speech being Irish the heart must needs be Irish for out of the abundance of the heart the tongue speaks." The complex discrepancies between the painfully acquired, civilized mother tongue of English and the pleasurable, tender, easy-to-consume, but nevertheless barbarous mother tongue of Irish shape Spenser's rich account of the conflict and collaboration between imperial domination and maternal nurture in *A View*.

Spenser's *A View* illuminates the intimate relationship between mother tongues and mothers' milk; it also highlights the increasingly central role that language and lactation come to play at this moment of national reformation. Richard Helgerson observes that Spenser's efforts to stamp out the barbarous dissonances of English—thereby rendering his native tongue capable of recapturing and even surpassing the linguistic power of classical Greek and Latin—suggest a partial turn from a dynastic conception of community, based on bloodline, to a post-dynastic, language-centered, nationalism.[80] *A View* also shows us that Spenser's anxieties

---

[78] Linda Gregerson, *The Reformation of the Subject*, p. 12.

[79] Edmund Spenser, *A View of the Present State of Ireland*, ed. W.L. Renwick (Oxford: Clarendon, 1970), p. 67.

[80] Richard Helgerson, *Forms of Nationhood: The Elizabethan Writing of England* (Chicago: University of Chicago Press, 1992).

about Irish wet nurses are closely related to this very same incipient shift from the old to the new nation, in which language and nurture, mother tongues and mothers' milk, are interlocking instruments of identity-formation and social belonging. *A View* endorses the idea that the nursing mother transmits the mother tongue of English to her offspring. But, even more provocatively, by connecting language and nursing with the heart ("out of the abundance of the heart the tongue speaks"), Spenser suggests that maternal nurture is the natural and affective source of English identity in the new linguistic nation.

Like his great teacher, Spenser, Milton attempts to rid English of barbarous dissonance and to render his mother tongue capable of surpassing Greek as the language of true liberty. Time and again in *Paradise Lost*, Milton identifies the reformation of language with the reformation of the nation, beginning with his prefatory note on "The Verse." Rhyme's "jingling sound," proclaims Milton, is "the Invention of a barbarous Age." In emancipating poetry from "from the troublesome and modern bondage of Riming," Milton meant his epic "to be esteem'd an example set, the first in *English,* of ancient liberty recover'd to Heroic Poem." As the language of "ancient liberty recover'd," English could achieve "true musical delight... [which] consists only in apt Numbers, fit quantity of Syllables, and the sense variously drawn out from one Verse to another." Released from the constraints imposed by rhyme and other false poetic ornaments, Milton's redeemed English verse emancipates the reader who is capable of appreciating its reformed harmonies. While the "neglect then of Rime" may seem "a defect... to vulgar Readers," to the refined, reformed reader, Milton's unrhymed verse drives off the "barbarous dissonance/Of *Bacchus* and his Revellers" (7.32–3) and thereby achieves both true liberty and "true musical delight."[81]

In the invocation to Book 7 of *Paradise Lost*, Milton signals the triumph of barbarism over poetry by alluding to the myth of Orpheus's dismemberment by the maenads, female Bacchantes who were jealous of the great poet's love for Eurydice. Not even Orpheus's mother, the Muse Calliope, was powerful enough to save her son from his horrible death: "nor could the Muse defend/Her Son" (7.37–8). But, in his note on "The Verse," Milton suggests he will do what Calliope could not: rescue Orpheus by emancipating the English language from the savage "Invention of a Barbarous Age." Inspired by his heavenly muse, who "with Eternal Wisdom didst converse" (7.9), the poet will surpass the power of Or-

---

[81] John Milton, "The Verse," in *The Complete Poetry and Essential Prose of John Milton*, ed. William Kerrigan, John Rumrich, and Stephen Fallon (New York: The Modern Library, 2007), p. 290.

pheus's mother, the classical muse of poetry. With its redeemed harmonies, Milton's unrhymed verse remembers Orpheus and restores music's "true... delight." As just noted, Cesare Ripa's emblem of Poesia as a lactating mother "exerted an inestimable influence on allegories of musical portraiture for over a century and a half."[82] It is possible to see something of Ripa's influential emblem of the lactating Poesia in "The Anonymous Life of Milton," written by Milton's student and amanuensis, Cyriack Skinner. The poet is said to complain to his amanuensis that: *"hee wanted to be milked"*:

> Hee rendered his Studies and various Works more early & pleasant by allotting them thir several portions of the day. Of these the time friendly to the Muses fell to his Poetry And hee waking early (as is the use of temperate men) had commonly a good Stock of Verses ready against his Amenuensis came; which if it happened to bee late than ordinary, hee would complain, saying *hee wanted to bee milked.*[83]

Not unlike Ripa's Poesia, the Milton in "The Anonymous Life" is depicted as bursting with poetic milk.

The poet describes himself through a similar maternal image of poetic plenitude in his prefatory note on "The Verse" in *Paradise Lost*. By pursuing "Things unattempted yet in Prose or Rhyme" in his epic, Milton represents himself as more powerful than Calliope, Orpheus's mother and the classical muse of poetry. Whereas Orpheus's mother could not save her son from dismemberment at the hands of the savage maenads, Milton as the new Poesia will rescue English verse from the "bondage of Riming," and emancipate the English reader from all such "custom[s]," which suppress "ancient liberty" and "true musical delight." Thus rather than force poetry into unity and coherence from without, Milton's blank verse will create coherence and harmony from within. While Orpheus's sweet verse was annihilated by uncontrollable, primitive, and female forms of barbarity, Milton's poetry will transmute the barbarous dissonances of English into civilized poetic harmonies capable of forging new inward forms of

---

[82] Austern, *Maternal Measures*, p. 241. Ripa's emblem might also be behind the speaker's depiction of his "speaking breast" (l.10) in Shakespeare's "Sonnet 23." Naomi J. Miller comments on the metaphor of nursing underlying this figure in "Playing 'the mother's part': Shakespeare's Sonnets and Early Modern Codes of Maternity," ed. James Schiffer (New York and London: Garland Books, 1999), esp. 347 and 355. Patrick Cheney comments on its theatrical implications in "'O let my books be... dumb presages': Poetry and Theatre in Shakespeare's sonnets," *SQ*, 52.4 (Summer 2001), esp. p. 244.

[83] Cyriack Skinner, "The Life of Mr. John Milton," *The Riverside Milton*, ed. Roy Flannagan (Boston: Houghton Mifflin, 1998), p. 12. In *John Milton: The Self and the World* (Lexington: University Press of Kentucky, 1993), p. 17, John Shawcross ascribes probable authorship of this brief "Life" to Milton's "former student, amanuensis, and friend Cyriack Skinner."

social connection—a new English poetry powerful enough to break down the natural bonds of blood and birthright, which organize the old dynastic nation. In theological terms, Milton's epic transforms the carnal Israel of the old dispensation into the spiritual Israel of the new era (a point to which we shall return in Chapter 4).

In addition to the nursing Poesia, Milton's self-image as a nursing poet/prophet recalls the aforementioned allegory of the minister as the breast of God in Puritan sermon literature. As we have seen, "Drinking the Word" emerges in the sermon as a comforting trope for the spiritual nourishment humankind could receive from the gospels, or God's "*Logical Milk*; 1 Pet.2.2," as described by the Puritan minister Samuel Willard. Fantasies of drinking God's Word as mother's milk replace the actual taste of the sacraments, the body and blood, on the tongue. "The whole World is a sucking Infant depending on the Breasts of Divine Providence," declared Willard.[84] As already noted, in *The Soules Vocation*, Thomas Hooker interprets Isaiah 66:11 with the same nursing fantasy in mind: "The Church is compared to a childe, and the brests are the promises of the Gospell; now the elect must suck out and be satisfied with it, and milke it out."[85] A primal form of spiritual hunger for the breast creates and sustains the bond between the Protestant believer and the paternal deity. This close connection between the paternal Word and mothers' milk obliquely acknowledges Christianity's great debt to Mary, which is otherwise occluded in Protestant worship.

Thus, no matter how morally vilified or scientifically disciplined out of existence, the mysteries of motherhood nevertheless endure, retaining at least part of their irresistible enchantment. As Spenser concedes, although barbarous, Irish nurse-milk nevertheless is "most pleasing."[86] In the next chapter, I look closely at the perverse allure and seductive danger associated with the figure of the bad or pre-rationalized mother. An extensive repertoire of anti-maternal stereotypes circulates in the period, most notably, the terrible Whore of Babylon and the malevolent Mother Nature. The regular circulation of these very familiar anti-female images apparently did little to detract from their shock value. Even when most overused, they find new life, as we shall see next.

---

[84] Samuel Willard, *A Compleat Body of Divinity* (Boston, 1726), pp. 32, 131.
[85] Thomas Hooker, *The Soules Vocation* (London, 1638), p. 306.
[86] Spenser, *A View*, p. 67.

2

Natural Mothers and the Changing "Character" of Englishness

A Pitiless Mother, Hic Mulier, and Macbeth

In *A Godly Form of Household Government*, Robert Cleaver and John Dod underscore the virtues of maternal breast-feeding by comparing "naturall mothers" to trees: "As therefore every tree doth cherish and nourish that which it bringeth forth, so also it becometh naturall mothers to nourish their children with their own milke."[1] Nature sets the standard by which the moral value of mothers can be measured. The character of the good mother is written in the bark, roots, and leaves of trees. Cleaver and Dod also appeal to nature to determine the differences between the good mother and bad one. Whereas the good mother models her conduct on nature's exemplary practices, the bad mother (falsely) values herself according to governing social customs, such as the aristocratic convention of hiring wet nurses. Cleaver and Dod dissociate the good mother from traditional measures of social rank by delivering their new moral equation between nursing and good mothering in the seemingly status-neutral terms of natural philosophy, specifically, botany, in the example just considered. Rather than represent the bias of a particular class or sect, the goodness of maternal nursing appears to reflect Nature's universal laws. In the guidebooks, the rhetoric of the new science measurably strengthens the Puritan guidebooks' critique of social custom and tradition: it also helps to explain the popularity of these oft-reprinted texts, which, as already noted, are aimed at readers from the middling classes.

As I shall demonstrate, a similar, but hitherto unexamined mixture of moral and new scientific rhetoric energizes and popularizes aspects of the *querelle des femmes*, the debate about the relationship between the two sexes. Not unlike the domestic guidebooks, the debate directed the period's ambivalences about the nature and value of "woman" to a popular

---

[1] Cleaver and Dod, 1621, sig. P4r.

middle-class readership.[2] Like the print sermon, the gender-debate pamphlets helped to mold an increasingly literate English population into an imagined commonwealth of engaged writers and readers, both male and female. As Sharon Achinstein observes, pamphlet writers "composed their audiences in supreme acts of fantasy, addressing their works to a public, demanding that their audiences read and respond to contemporary issues; they also presented models for public debate by fighting pen-battles in print."[3] By making the nature and value of women the shared concern of a fantasy public audience of responsive readers and writers, (a popular republic of letters), the print debate about the two sexes provides a material record of the conceptual shift from the corporealist model of the nation as *natio* (bound together by blood, soil, and kinship bonds) to the abstract paradigm of the modern nation (unified through affective and linguistic modes of affiliation).

The debate also illuminates how this national paradigm-shift is gendered. As Henderson and McManus argue, the gender debate helped to give women new authority, self-definition, and public access: "for the first time in England women began to write in their own defenses and for the first time anywhere significant numbers of women began to publish defenses."[4] The stock-character names of some of the best known of these writers, such as Esther Sowerman, Jane Anger, Constantia Mundi, and Joan Hit-him-home, suggest that these might well be pseudonyms adopted by male authors.[5] But, whatever the actual gender of its participants, the gender debate did successfully *stage* a conversation "amongst

[2] As Katherine Usher Henderson and Barbara F. McManus note: "during the Renaissance the English middle class had a distinct taste for this fare."Henderson and McManus, "The Contexts," *Half Humankind: Contexts and Texts of the Controversy about Women in England, 1540–1640* (Urbana and Chicago: University of Illinois Press, 1985), p. 3. Quotations from *A pitiless Mother, Hic Mulier* and *Haec Vir* are taken from this volume and noted in the text.

[3] Sharon Achinstein, *Milton and the Revolutionary Reader* (Princeton: Princeton University Press, 1994), p. 110.

[4] Henderson and McManus argue that "while definite proof either way is lacking, probability strongly supports our contention that these writers were indeed women, as they claimed," "Female Authorship," *Half Humankind*, p 21. In "Crossdressing, the Theatre, and Gender Struggle in Early Modern England," *Shakespeare Quarterly* 39 (1988): 427–8, Jean Howard argues that the invention of printing did not simply increase "the ways in which women could be controlled and interpellated as good subjects of a patriarchal order." Access to print also allowed women "to rewrite their inscriptions within patriarchy."

[5] In "Muzzling the Competition: Rachel Speght and the Economics of Print," in *Debating Gender in Early Modern England, 1500–1700*, p. 59, Lisa J. Schnell observes that Rachel Speght was "the one writer in the controversy who was inarguably a woman." In "Material Girls: The Seventeenth-Century Woman Debate," in *Women, Texts and Histories 1575–1760*, ed. Clare Brant and Diane Purkiss (London: Routledge, 1992), Diane Purkiss argues against reading the debate as an expression of early modern feminism.

friends where the number of each sexe were equall," as Esther Sowerman writes in *Ester hath hang'd Haman*.[6] The anonymous author of *Hic Mulier*, for example, is acutely aware that he is participating in at least the pretense of an equitable conversation between men and women. He not only directly addresses the noblewomen whose transvestite costumes he decries, but he also impersonates them, rehearsing their rebuttal to his attacks on cross-dressing in their collective voice: "But now methinks I hear the witty offending great Ones reply in excuse of their deformities: 'What is there no difference among women?'" (272). The print debate about gender thus does a good deal of cultural work. It provokes readers of both sexes into disputatious print conversation with one another—to publicly proclaim their personal opinions rather than silently endorse officially formulated truths.

The gender debate finds its elite origin in manuscript circulation on the Continent in medieval-and-early-Renaissance France and Italy, in texts such as Boccaccio's *Concerning Famous Women*, Christine de Pizan's *The Book of the City of Ladies*, and Cornelius Agrippa's *The Declamation on the Nobility and Preeminence of the Female Sex*. In Jacobean England, the debate shifts from manuscript to print and acquires new popularity, especially through the Swetman controversy, which centered on Joseph Swetman's provocative printed pamphlet, *The Arraignment of Lewd, idle forward, and unconstant women*, and the three pamphlets that refute Swetman's anti-female arguments.

One objective of the larger debate, which continued into the 1640s, was to characterize the good mother. I focus specifically on two cheap-print, gender-debate pamphlets that associate the good mother with the nursing mother: *A pitilesse Mother*, a reportorial pamphlet published in 1616; and *Hic Mulier*, a 1620 pamphlet decrying female cross-dressing. As in the guidebooks, in these two pamphlets, the character of the good mother emerges from the victory of reason and science over custom and tradition, which inevitably fail to withstand rationalist or empiricist critique. The good woman in *Hic Mulier* resembles the mother of the guidebooks: she is the *natural* mother who nurses her own children, physically and/or spiritually. As in the guidebooks, the character of the good mother in the two pamphlets is unaffected by class, social rank, or privilege. She is a natural kind rather than a social type or "Character," in the Theophrastic terms that are exemplified by the character-writings of Joseph Hall, Thomas Overbury, and John Earle, a genre to which we shall soon return.

---

[6] Esther Sowerman, *Ester hath hang'd Haman or An Answer to a lewd Pamphlet, entitled The Arraignment of Women* (STC/22974, London, 1617), sig. A2r.

The appeal to reason over custom is especially vivid in the pro-female pamphlet *Haec Vir*, a "briefe Dialogue between *Haec Vir the Womanish Man and Hic Mulier the Man-Woman*" (277). Hic Mulier challenges the charges made against her in *Hic Mulier*, the pamphlet: that she is monstrous, barbarous, perverse, and so forth. Notably, in *Haec Vir*, Hic Mulier defends herself by deploying the very same mixture of rationalist and empiricist rhetoric that the author of *Hic Mulier* utilizes to prove that female transvestitism is unnatural. Hic Mulier (the character in *Haec Vir*) wrests control of reason from the author of *Hic Mulier* in order to make the counter-argument that, by defaming women cross-dressers, the *Hic-Mulier*-author merely defends custom, and that hence it is he, not she, who is completely unreasonable: "To conclude, Custom is an Idiot, and whosoever dependeth wholly upon him without the discourse of Reason will take from him his pied coat and become a slave indeed to contempt and censure" (284). As we shall see, as in *Haec Vir*, arguments for the preeminence of reason over custom, and vice versa, structure the pro-or-con assessments of women represented in the *querelle des femmes*.

The guidebooks' and gender-debate pamphlets' attempts to cure character flaws by deploying reason and empirical observation against custom markedly differ from traditional humoral categorizations of personality or temperament. As already noted in my first chapter, within the governing Galenic framework, character, or temperament, is a function of the four humors. Although both the gender-debate and guidebook writers continue to understand physical health and character or temperament in humoral terms, they also appeal to the new philosophy's emphasis on directly observed experience to characterize human behavior. Unlike temperament, which results from mixtures governed by a dominant humor for which no empirical evidence could be summoned, character in the guidebooks and debate pamphlets is empirically understood, translated into natural-and-civic law, and then put into social practice—part of a larger disciplinary Protestant project for elevating the moral character of the English nation as a whole. Unlike traditional humoral definitions of character, which maintain the status quo, new scientific, legal, and rationalist methods for classifying moral character tend to challenge customary values.

The construction of character in the guidebooks and debate-pamphlets also differs in striking ways from the aforementioned literary genre of character-writing, which rode a wave of popularity in the first two decades of the seventeenth century. Overburian character-writing takes its inspiration from Theophrastus, a disciple of Aristotle, who is generally considered the father of botany. Theophrastus is the author of the satiric *Moral Characters*, translated from the Greek into Latin as *Characteres*

Ethicae Theophrasti, and first translated into English by Joseph Healey in 1628. Under the title, *Theophrastus His Morall Characters* or *Description of Maners*, Healy's translation was published with John Earle's *Microcosmography or, A Piece of the World Discovered in Essays and Characters* (1628). An Anglican and a Royalist, Earle was appointed chaplain and tutor to the future Charles II in 1641. In 1643 he was elected to the Westminster Assembly, a position he declined because of his sympathies with the Anglican Communion and Charles I.[7] Also in 1643 he was elected chancellor of Salisbury Cathedral, but deprived of this position as a malignant. After the Restoration, in 1663 he was consecrated as Bishop of Salisbury. The earliest example of Theophrastic character-writing in English is Joseph Hall's *Characters of Vice and Virtue*, published in 1608. It is worth noting that 1608 also is the year of John Milton's birth. Milton later would demolish Hall polemically in *An Apology to Smectymnuus*. Like Hall and Earle, Thomas Overbury was staunchly anti-Puritan. He characterizes "the Puritane" as a "diseased piece of Apochrypha": "his greatest care is to contemn obedience, his last care to serve God handsomely and cleanly…the fumes of his ambition make his very soul reel."[8]

Overbury's *Characters* derives some of its fame from the author's role in one of the most sensational scandals of the period. The other two main players were Frances Howard and Robert Carr, who was Overbury's close friend. In 1612 Rochester began an affair with Howard, which Overbury vehemently opposed. At this same time, Overbury wrote his poem *A Wife*, which circulated widely in manuscript. Howard soon learned that Overbury had written this poem to open Rochester's eyes to the defects of her character. She retaliated by manipulating King James into imprisoning Overbury in the Tower on trumped-up charges on 22 April 1613; she also had the king annul her marriage to Devereux, so that she could be free to marry Rochester. Overbury remained in the Tower until September 1615, when it was discovered that he had been poisoned. The prime suspects were Rochester and Frances Howard, who were both executed. Edward Coke and Francis Bacon later determined that Howard with the assistance of a Mrs. Turner had been feeding Overbury sulfuric acid.[9]

Among other things, the scandal provided fodder for the popular gender-debate. Howard's infamous conduct sparked new volleys of pro- and anti-female pamphlets, including *Hic Mulier* and *Haec Vir*. We shall soon see that Howard and Mrs. Turner are the actual targets of the *Hic*

---

[7] John Spurr, "Earle, John, (1598x1601–1665)", *ODNB*, online edn, January 2008 (http://www.oxforddnb.com/view/article/8400).

[8] Thomas Overbury, *New and Choise Characters* (STC/1732:25, London 1615), sig. E1r

[9] For a full account of the scandal, see David Lindley, *The Trials of Frances Howard: Fact and Fiction at the Court of King James* (New York: Routledge, 1993).

Mulier-author's repudiation of female transvestites. My broader aim at this juncture, however, is to underscore that, in addition to character-izing specific social types, as do Overbury, Hall, and Earle, the anti-female authors of *Hic Mulier* and *A pitiless Mother* (like the pro-maternal breast-feeding Puritan guidebook authors) also attempt to reform "woman" by redefining her virtues in natural terms, most susceptible to rational-and/or-experimentalist analysis. Whereas Overbury and other character-writers create fixed social types, the authors of the two gen-der-debate pamphlets attempt to improve women's social character.

Macbeth is the third text that I consider in this chapter. Surprisingly scant attention has yet been paid to the relationships between the *querelle des femmes* and Shakespeare's play. I argue that reading Shakespeare's play through the lens of the gender-debate opens a new window on *Macbeth*'s much-discussed maternal matters. In *Macbeth*, as in *A pitiless Mother* and *Hic Mulier*, the bad woman has bad breasts.[10] Rather than follow nature by breast-feeding her own children, Lady Macbeth would rip the babe from her breast and dash its brains out to prove her male mettle.[11] In this, Lady Macbeth resembles the women transvestites in *Hic Mulier*: these cross-dressed noblewomen also defy nature by artfully making their bad, sexu-ally provocative, bare breasts into the centerpiece of their *haute couture* male attire. As I shall demonstrate, in Shakespeare's play as in the two pamphlets, good maternal nurture represents the antidote to mounting tensions about shifting national borders, generated partly by James I's vision of an integrated Scotland and England. Anxieties about changing national boundaries and identity are redirected, with a special focus on the breast, toward the "monstrous" union of the two genders, as exemplified

---

[10] On the distinction between the good breast and the bad one, see Melanie Klein, "The Psychogenesis of Manic-Depressive States," in *The Selected Melanie Klein*, ed. Juliet Mitch-ell (New York: Free Press, 1986), p. 141: "The child's libidinal fixation to the breast devel-ops into feelings for [the mother] as a person. Thus feelings both of a destructive and of a loving nature are experienced toward one and the same object, and this gives rise to deep and disturbing conflicts in the child's mind." As a "defense against anxiety," the child splits the breast into two: a "good" breast and a "bad" breast. These two opposing attitudes to-wards the breast coexist latently without resolution within the realm of infantile fantasy, ultimately evolving into and manifesting themselves as ambivalence about other kinds of integrated objects.

[11] Mary Floyd-Wilson in "English Epicures and Scottish Witches," *SQ*, 57.2 (Summer 2006): 136–7 observes that, traditionally, Scottish women were strongly encouraged to nurse their own babies, since mothers' milk was thought to encourage heartiness and to prevent degeneracy. Floyd-Wilson relates Lady Macbeth's disdain for maternal breast-feeding to early modern Scottish perceptions that their nation's character was beginning to lose its "auld" ruggedness. Stephanie Chamberlen argues that Lady Macbeth's infanticidal fantasy invokes a maternal agency that momentarily empowers her illegitimate political ambitions. See "Fanta-sizing Infanticide: Lady Macbeth and the Murdering Mother in Early Modern England," *College Literature*, 82.3 (2005): 72.

by Lady Macbeth and the female transvestites of *Hic Mulier*. Unlike the two pamphlets, which challenge custom through reason, however, *Macbeth* aligns the consolidating national power that symbolically accrues to the good breast with the traditional interests of dynastic government and the old sacramental monarchical order. Not unlike the gender-debate, however, *Macbeth* does not completely foreclose on the proto-modern national community of intellectually and emotionally engaged readers and writers, imagined by the new print culture.[12] Rather the play makes room for the new ethically self-directed and disputatious national subject to emerge—a space clearing that complicates *Macbeth*'s alignment of re-enchanted maternal nurture with the regenerative magic of English kingship, encapsulated by Malcolm's hagiographical depiction of Edward II. To best appreciate *Macbeth*'s complex ideological negotiations, however, we need first to say more about the relations between the new mother and the reformed nation in the two pamphlets.

NURSING NIGHTMARES AND NATURAL ENGLISHNESS IN *A PITILESS MOTHER*

A pitiless Mother is a tabloid-like news account of Margaret Vincent, an English gentlewoman of "good parentage," who after falling "into the hands of Roman Wolves," proceeded to murder her two youngest children in a misguided act of religious enthusiasm on Holy Thursday.[13] The full title of the pamphlet, *A pitiless Mother That most unnaturally at one time murdered two of her own Children at Acton upon holy Thursday last, 1616*...emphasizes Margaret's unnaturalness—a term, which, as Greenblatt documents (and as discussed in my first chapter), takes the place of the profane in the early modern imaginary. Margaret's Roman influence transforms her into "a fierce and bloody Medea": she hangs her children "with a Garter taken from her leg" (364)—the "Garter" here is literally exposed as a murder weapon of mythic proportions. The pamphlet-author recounts that "to save their souls (as she vainly thought) she proposed to become a Tigerous Mother and so wolfishly to committ the murder of her

---

[12] Elizabeth Eisenstein analyzes the ways in which print culture helped to change the interpretive habits of readers in her landmark study, *The Printing Press as an Agent of Change* (Cambridge: Cambridge University Press, 1985). See also Achinstein, "Revolution in Print: Lilburne's Jury, *Areopagitica*, and the Conscientious Public," *Milton and the Revolutionary Reader*, Chapter 1.

[13] Naomi J. Miller reads *A pitiless Mother* as a cautionary tale condemning the tragic fate of women who challenge the authority of their husbands in *Maternal Measures: Figuring Caregiving in the Early Modern Period*, ed. Naomi J. Miller and Naomi Yavneh (Burlington, VT: Ashgate, 2000), p. 7.

own flesh and blood, in which opinion she steadfastly continued, never relenting *according to nature*" (363, emphasis added). Afflicted by "that infectious burden of Romish opinions" (362), Margaret exchanges her natural desire to nourish her children for the unnatural violence that always lurks just behind the surface of the idealized act of nursing—much as Lady Macbeth takes her milk for gall and her tender babe for a bloody child. Brutal rather than loving, "Romish" rather than English, Margaret's perverse maternity is "more unnatural than Pagan, Cannibal, Savage, Beast, or Fowl" (367). Encompassing the heretical, the primitive, the taboo, and the bestial, Margaret's unnatural motherhood threatens to overwhelm the tiny natural island that is civilized England.

Rather than advance English civilization through the natural expression of maternal nurture, Margaret first succumbs to and finally is destroyed by an unnatural, cannibalistic, Catholic hunger that tears away at the nation's civilized structures of family and state. Her desire to murder her children is attributed to an insatiable, primitive thirst for blood, which ultimately proves to be self-destructive:

> These two pretty children being thus murdered without all hope of recovery, she began to grow desperate and still to desire more and more blood…so taking the same garter that was the instrument of their deaths, and putting the noose thereof about her own neck, she strove therewith to have strangled herself (363)

Made crimson by the blood of her innocent children, Margaret's garter and her terrible Catholic passions mark her as the scarlet prostitute, the Whore of Babylon. In murdering her children she performs a perverse form of Catholic ritual, which makes manifest the violent act of eating and drinking implied by taking the sacraments. Margaret forces her children to drink from her "bitter cup," her poisoned chalice. After hanging her infant, she compels her elder child to "drink of the same bitter cup as she had done the other" (363) and then she ritually drinks from "the same bitter cup" herself. Margaret's maternal unnaturalness exposes the literally fatal influence of the old Church's rites and rituals. Unlike the malevolent Roman mother, the benign Protestant mother is a natural nurse. She resembles the self-sacrificing, Christ-like pelican rather than the evil viper or satanic serpent:

> By nature, Margaret should have cherished them with her own body, as the Pelican that pecks her own breast to feed her young one with her blood, but she, more cruel than the Viper, the envenomed Serpent, the Snake, or any Beast whatsoever, against all kind that takes away those lives to whom she first gave life (364)

As the quick slide from "the Pelican" to "the envenomed Serpent" (364), from redemptive blood to poisonous venom, suggests, Margaret's descent into unnatural Roman motherhood swiftly recapitulates the larger narrative of scriptural history from the Fall to the Crucifixion, except in *reverse.*

Margaret's loss of proper maternal feeling is the chief symptom of her unnaturalness as a mother. Time and again, the pamphlet-author narrates that Margaret's "heart [was] without all motherly pity." Her pitilessness exposes not only her savage unnaturalness, but her foreignness as well: even "the Cannibals that eat one another, will spare the fruits of their own bodies, the Savages do the like.... Yea, every beast and fowl hath a feeling of nature and according to kind will cherish their young ones" (367). Devoid of the natural desire to cherish and nurture her offspring, Margaret is driven by perverse maternal impulses: "So tyrannous was [Margaret's] heart" that she is unmoved by the affecting face of her own child: "a countenance so sweet that might have begged mercy at a tyrant's hand" (364). Margaret's "tyrannous... heart" is so central to the pamphlet's narrative and, as we shall see, to its nationalist concerns that it becomes a plot point in an even larger tale of natural maternal nursing gone awry. Having been "nursed by the Roman sect," Margaret herself becomes a brutal Roman Catholic mother who not only murders her children instead of nourishing them, but who also insists that murder *is* nurture: "she proposed to become a Tigerous Mother and so wolfishly to committ the murder of her own flesh and blood, in which opinion she steadfastly continued" (362). In this nursing nightmare, Roman milk brutally suppresses the very meaning of "meaning," eradicating normative distinctions between caring and killing, life and death. The bad maternal breast, associated with the ubiquitous image of the Roman Whore, is the portal to a dark, relativist, alien or "other" world, in which normative, quotidian experience is transformed into a nightmarish dreamscape, and where, as for the tortured tyrant, Macbeth: "Nothing is but what is not" (1.3.141).

Whereas Margaret's crimes establish Roman milk as the foreign source of her uncharitable tyranny, and pitilessness, the temperate and charitable justice that she is served "according to law" (366) forms a metonym for Mother England's nursing powers. The clear, measured language of English law overrides the discursive chaos and crisis of meaning and identity that the tyranny of Roman nursing engenders. The "subtle sophistry of some close Papists" binds Margaret to "a belief in bewitching heresy"; in contrast, the rational, disenchanted discourse of English law and justice, epitomized by "the *Mittimus*," or warrant (366), which commits Margaret to the keeping of a jailor in Newgate prison, clears the English family, language, and national landscape of the primitive maternal unruliness unleashed by Catholicism's bewitching heresies, deadly rituals, and unnatural mothers' milk.

CROSS-DRESSED WOMEN AND NATURAL MOTHERS: BOUNDARY PANIC IN *HIC MULIER*

Not unlike *A pitilesse Mother*, *Hic Mulier* exposes the breast as an inherently unstable cultural signifier—a mark, on the one hand, of socially productive feeling, of national belonging, and normative or natural identity, and, on the other hand, the source of unnatural and hence uncivilized and so non-English thought, emotion, and action. The socially disruptive bad breast is the focus of the *Hic Mulier*-author's attention in his diatribe against women transvestites. The cross-dressed costume consists of a "ruffian broad-brimmed hat and wanton feather...the loose lascivious civil embracement of a French doublet, being all unbuttoned to entice (or 'to reveal naked breasts,' as Stephen Orgel glosses this), most ruffian short hair," and a sword (267). Orgel argues that, despite her transgressive display of conspicuously male attire, her hat, her doublet (a masculine garment), her short hair, and her sword, the woman transvestite's aim is not to pass as a man.[14] Rather, by baring her breasts and carrying a sword, the female cross-dresser proclaims that the erotic potential of the feminine is the same or greater than that of the masculine. Orgel's apt point is that, by cross-dressing, early modern English women advertised their sexual freedom—an assertion of male autonomy that, drawing on the commonplace association of female cross-dressing with prostitution, the *Hic Mulier*-author associates with whorishness: they had "laid aside the bashfulnesse of [their] nature to gather the impudence of Harlots" (266). For Orgel, the new anxiety about female cross-dressing that surfaces at the beginning of the seventeenth century stems from the woman transvestite's usurpation of male sexual-and-political power—an anxiety also articulated in King James's admonition to the London clergy in 1620, that they "inveigh vehemently and bitterly in their sermons against the insolency of our women, and their wearing of broad-brimmed hats, pointed doublets, their hair cut short or shorn, and some of them stilettos or poniards."[15]

As Orgel suggests, the dangerous bare-breasted transvestite in *Hic Mulier* reflects a new unease about female sexual license and political

---

[14] Ann Rosalind Jones and Peter Stallybrass read *Hic Mulier* as focused on "women's regendering of the head, the very symbol of patriarchal authority": "The man-woman of the 1610s emerges less from the hybridization of her lower parts (skirts/breeches) than from hybridization of the head," in *Renaissance Clothing and the Materials of Memory* (Cambridge: Cambridge University Press, 2000), p. 78.

[15] Stephen Orgel, *Impersonations: The Performance of Gender in Shakespeare's England* (Cambridge: Cambridge University Press, 1996), p. 119; Angeline Goreau, *The Whole Duty of a Woman: Female Writers in Seventeenth-Century England* (Garden City, NY: Doubleday, 1985), p. 91.

power partly due to the large shadow that the figure of the late Elizabeth I cast across the Jacobean court. Pointing Orgel's intriguing observation in a somewhat different direction, I argue that the new unease about the sexually hybrid figure of the female transvestite conveys not only cultural anxieties about female sexual power, but also the crisis of Englishness that results from early modernity's shifting paradigms of the nation. In political terms, this national identity crisis was exacerbated in good part by James's campaign to federate Scotland with England. As already discussed in my Introduction, Union threatened to dissolve enduring ethnic and religious differences between the two peoples and undermine England's sense of itself as the not-Scotland. In the pamphlet, the natural mother, with her consolidating, civilizing, good breast, prevents the potential dispersal of customary personal and national identity attributed to Union. She is the gatekeeper between what is English and what is not. The new anxiety that Orgel detects in *Hic Mulier* thus represents a response not only to the female sexual license and independence asserted by bare-breasted women in men's clothes, but also to the exotic doubleness of the transvestite's gender, and the threat that s/he presents to the integrity and intactness of the traditional nation and national subject.

As just noted, *Hic Mulier* works out its overlapping preoccupations with national identity and maternal nurture in relation to two specific women, Anne Turner and Frances Howard, and the spectacular court scandal in which they both played leading roles. The *Hic Mulier*-author never mentions the two women by name. He distinguishes each, "the one ... the other," only by her fashion choices. Mrs. Turner, "the one cut from the Commonwealth at the Gallows," is condemned for introducing the fashion of wearing a yellow ruff and cuffs into the Jacobean court: "From the first you got the false armory of yellow Starch" (267). As Henderson and McManus point out, this was "a fashion that James despised" (267). Mrs. Turner's predilection for clothes starched with yellow unambiguously stamps her as vulgar and base: "for to wear yellow on white or white upon yellow is by the rules of Heraldry baseness, bastardie, and indignitie." By contrast, Howard is specifically identified with the transvestite's unnatural male–female, bare-breasted attire:

> From the other you have taken the monstrousness of your deformity in apparel, exchanging the modest attire of the comely Hood, Cowl, Coif, handsome Dress or Kerchief, to the cloudy Ruffianly broad-brimmed Hat and wanton Feather, the modest upper parts of a concealing straight gown, to the loose lascivious embracement of a French doublet, being all unbuttoned to entice (267)

This sensationalized depiction of Howard with short hair, naked breasts, and a French doublet differs in striking ways from the complimentary images rendered in contemporary court portraits of the Countess, such as the engravings made by Simon van de Passe in the 1610s and early 1620s. The *Hic Mulier*-author emphasizes the monstrousness of Howard's exposed breasts, which protrude out of her French doublet, "being all unbuttoned to entice" (267), while van de Passe's engravings depict the bare-breasted, short-haired Countess as a paragon of courtly fashionableness. As Ann Rosalind Jones and Peter Stallybrass observe, these engravings feature "the almost full exposure of her breasts, not…[as] a sign of her particular wantonness but the common court fashion of the mid-1610s."[16] In the van de Passe engravings, Howard's short hair and exposed breasts serve as stylish markers of her elite status and currency at court. In *Hic Mulier*, however, the same cross-dressed costume and exposed breasts inscribe the Countess not as a distinctively aristocratic figure, but as a general anti-female type, the bad woman—the same general category into which the pamphlet places the seamstress Turner, Howard's partner in (fashion) crime.

As I have just noted above, although the pamphlet-author differentiates the two women by their styles of dress, he never directly names Turner and Howard; nor does he mention the differences in their social status. He repeatedly underscores that both women are naturally perverse. Their fashion crimes, although different, render them equally monstrous. While the specific nature of each woman's fashion crime is duly noted, in every other way, the pamphlet chooses to paint Turner and Howard as categorically evil. As we shall see, unlike the van de Passe engravings, which offer a very different kind of social coding, *Hic Mulier*'s representations of the cross-dressed Howard reflect the larger cultural and ethical ambitions of the gender-debate: to produce a broad, moral category of "woman" comprehensible to the social totality.

Both Howard and Turner were found guilty of poisoning Overbury, but while Howard's sentence was later commuted by the king, Turner was hanged in November 1615 wearing, at James's insistence, a dress, cuffs, and a ruff dyed with yellow starch. This is a notable fashion detail. As just noted, yellow was customarily associated with illegitimacy, but, as Jones and Stallybrass observe, yellow, the color of both saffron dye and urine, newly "linked luxury and contaminating waste" in the early modern cultural imagination. Yellow clothing was seen "as originating with England's traditional enemies, Spain, Ireland, and France, all seats of 'the Whore of Babylon.'"[17] By wearing yellow, Turner, a Catholic, and Howard, of the

---

[16] Jones and Stallybrass, p. 79.

[17] Jones and Stallybrass, p. 67. See *Renaissance Clothing and the Materials of Memory*, Chapter 3, for a fascinating discussion of the Jacobean fashion of yellow starch.

crypto-Catholic Howard family, strike an insurgent Duessa-like pose, not unlike the Empress in Dekker's *The Whore of Babylon*. James's insistence that Turner should be hanged in her yellow cuffs sends a message to all those who like Turner and Howard would make the contaminating influence of Catholicism and foreign luxury seem fashionable and appealing.

Similar connections between foreignness and *haute-couture* fashion can be found in the writings of John Bulwer, the seventeenth-century natural philosopher. For Bulwer, fashionable English dress replicates the exotic markings (body painting, scarring, circumcision, and piercing) on foreign bodies. As Stephen Greenblatt points out, Bulwer argues that "contemporary English clothing at its most fashionable actually reproduces many of the transformations that are carried out in other cultures on the flesh itself."[18] For Bulwer, high fashion offers a way to impersonate or literally to in-habit the Other: to "go native," as it were. The *Hic Mulier*-author cautions against this seductive primitivism in his title through the classical Latin verse that he adapts from Virgil's *Ecologue* 8.63: "*Non omnes possumus omnes.*" For Bulwer, the doublet, a man's garment, which is also mentioned several times in *Hic Mulier* as the frame for the female transvestite's bare breasts, offers an especially clear example of the strange likeness between the fashionable cut of a well-tailored garment and the decorative body cuttings of primitive peoples. Bulwer writes: "the slashing, pinking, and cutting of our Doublets, is but the same phansie and affection with those barbarous Gallants who slash and carbonado their bodies."[19] Despite the differences between Bulwer's ethnography and the *Hic Mulier*-author's gender polemic, both writers share the same antipathy toward ornament, decoration, and ritual, whether Catholic or primitive.

In *Hic Mulier*, women thus are exhorted to "Imitate Nature," rather than decorate or mutilate, scar, pierce, and tattoo themselves in the name of high fashion. Women in male attire are "most monstrous"; they have made their bodies into "not half man/half woman, half fish/half flesh, half beast/half Monster, but all odious" (266). They are the "stinking vapors drawn from dunghills," "living graves, unwholesome Sinks, quartan Fevers for intolerable cumber, and the extreme injury and wrong" (269). Their "deformity" offends "man in the example and God in the most unnatural use." They "mould their bodies to every deformed fashion" and "lose all charms of women's natural perfections." If, in moral terms, female cross-dressing is categorically bad, it also is a social pathology: a "quartan

---

[18] Stephen Greenblatt, "Mutilation as Meaning," in *The Body in Parts*, p. 236.
[19] John Bulwer, *Anthropometamorphosis: Man Tranform'd*, 2nd ed. (London: J. Hardesty, 1653), quoted in Greenblatt, "Mutilation and Meaning," p. 235.

Fever" and "an infection that emulates the plague" (269). The unnatural, unhealthy body is also a foreign body. Cross-dressing is "all base, all barbarous"; it is "barbarous, in that it is exorbitant from Nature and an Antithesis to kind, going astray with ill-favoured affectation both in attire, in speech, in manners" (268). As in Bulwer, in *Hic Mulier*, the marks of culture, such as dress, speech, and manners ("[transvestites] mould their bodies to every deformed fashion, their tongues to vile and horrible profanations, and their hands to ruffianly and uncivil actions"), are to be understood as natural attributes. High fashion slides into physical deformity, political unruliness, and foreign contagion: these are overlapping symptoms of a much larger national infection that eats away at native or traditional Englishness. Cross-dressing, writes the *Hic Mulier*-author, "is an infection that...throws itself amongst women of all degrees, all deserts, and all ages; from the Capitol to the Cottage are some spots or swellings of this disease." In its indifference to degree and other marks of social distinction, female cross-dressing is an all-encompassing social illness, affecting the integrity of the national whole, "from the Capitol to the Cottage" (269). Thus, whereas "Good women" are "Seminaries of propagation," who "maintain the world...and give life to society," bad women, as epitomized by the female transvestite, spread devastating social epidemics that deprive the nation of its health and vitality.

Whereas in *A pitiless Mother* English law represents the antidote to the enchanted dogma that Margaret Vincent imbibes at the breast of her Roman nurse, in *Hic Mulier*, medicine serves as the civilizing antidote to the female cross-dresser's licentious deformation of the English character. The title of the pamphlet announces the medical concerns of its "brief Declamation": *"A Medicine to cure the Coltish Disease of the Staggers in the Masculine-Feminines of our Times."* The proposed cure is a change of dress. The antidote to the plague-like national infection generated and transmitted by women in men's clothes is the plainly dressed figure of the good woman celebrated by the pamphlet. Whereas the female transvestite infects the national body with "spots and swellings" (269), the properly attired, plain-styled woman is associated with social and spiritual health: she represents "a world full of holy thoughts" (271). In *Hic Mulier*, women's dress and bodies, most especially the breast, become the focus of Protestantism's anti-sacramental and anti-customary social energies.[20] Although uncovered, the transvestite's "bared breasts seducing" (271) are part of a costume that ornamentally conceals the natural corruption and categorical badness of women in male clothes. In contrast, the good

---

[20] Joan Webber, *The Eloquent "I": Style and Self in Seventeenth-Century Prose* (Madison: University of Wisconsin Press, 1968).

woman exemplifies the new dressed-down, plain-styled virtues by concealing her breasts: "oh hide them, for shame, hide them in the closest prisons of your strictest government" (271).

In *Hic Mulier*, then, the breast registers the same paradoxical relationship between female erasure and presence, exclusion and inclusion that characterizes the period's larger revaluation of maternal nurture and nature. By associating female virtue with "sober shows without" (271), the pamphlet tries to suppress the transvestite's assertion of sexual independence and social power and presence. Unlike bad women, good women have no distinctive, external markings by which their socioeconomic worth can be measured. Like the domestic guidebooks, *Hic Mulier* writes women out of the public realm by abstracting female worth into incalculable, wholly inward terms: "holy thoughts" (271). The cross-dressed women represented in the pamphlet resist this kind of abstraction, public erasure, and private enclosure by usurping male power through their sexually and socially provocative dress: their swords and naked breasts. But, while peripheral in a social universe where birth, blood, and soil are the ruling measures of social value, the natural mother gains prestige and significance as a consolidating agent for the emergent imagined community that the new print culture helps to usher into modern existence. Whereas the female transvestite challenges the idea that "we all be coheirs of one honor, one estate, and one habit," good women in their chaste uniforms represent the oneness of honor, estate, and habit—the same uniformity that the hybrid male–female transvestite undermines. In good women are "all harmonies of life, the perfection of Symmetry, the true and curious consent of the most fairest colors" implanted by God "in your first creation" (272). The plain-styled woman is rewarded for her unadorned virtue, especially the "honest care" of her children, with the highest expressions of undying love and the bonds she makes with men's souls: "you shall draw men's souls unto you with that severe, devout, and holy adoration, that you shall never want praise, never love, never reverence" (272). Denied the erotic breast, she is granted the spiritual breast: the focal point of the new nurturing mother. While the artful female transvestite transmutes maternal love into enticing but deadly forms of psychic, spiritual, and social disease, the plain-styled, natural mother converts unruly and unhealthy, erotic desire into disciplined and salvific forms of maternal love. The good, natural mother provides access to a redemptive "world of holy thoughts" (271) and a new point of pure origin: "your first creation" (272).

Richard Wilson's analysis of the petition for a midwives' charter presented to James I in 1616 helps to illuminate more precisely what is at stake politically in *Hic Mulier*'s distinctions between natural mothers and cross-dressed women: between the good breast and the bad one. For Wilson, the midwives' petition represents the material cause for the conceptual translation of the old maternity into the new—a rewriting that would bring conception and birth out of the untutored, female birth room and into the professional chamber of the physician and, finally, up to the halls of enlightened government. Drafted by the London midwives themselves, this petition was promoted by the brothers, Peter Chamberlen, the elder, and Peter Chamberlen, the younger, who invented the obstetrical forceps. The Chamberlen brothers' professed aim in promoting the licensing of midwives was to avoid the dire "consequence for the health and strength of the whole nation if ignorant women, whom poverty or the fame of Venus hath intruded into midwifery, should be insufficiently instructed."[21] As Wilson points out, however, the unspoken ambition behind the Chamberlens' support for the midwives' petition was less improvement of the material conditions and practices of birthing than "the paternal licensing of fertility itself."[22] In the 1670s Hugo Chamberlen (grandson of Peter, the younger) tried to sell the secret of the iron forceps to the French.[23]

A similar political ambition can be detected in James I's self-representation as a "nourish king." In *Basilikon Doron*, James reminds Prince Henry that monarchy displays both fatherly power and motherly nurturing: "it [is] one of your fairest styles to be called a louing nourish father" who provides his subjects with "nourish milke."[24] I shall discuss this trope and *Basilikon Doron* at greater length in the next chapter, but nevertheless I would like to note here that James's highly motivated appropriation of maternal nurture is crucial to his campaign to sell his Unionist idea of Great Britain to Parliament. As already suggested in my Introduction, resistance to Union was made in the name of Mother Country, that of Elizabethan England rather than Jacobean Britain or Albion. Richard Perceval's aforementioned speech on this subject to Parliament is worth repeating here: "the name of our Mother Country to be kept: Our desire natural and honorable—she hathe nurst, bred, and brought us up to be men."[25] Perceval's remarks place the received trope of England as maternal nurse in service of an imagined English insularity. Partly to wrest control over this affecting anti-Unionist equation between nation and lactation—and the natural desires from which it derives

---

[21] Quoted in Wilson, "Observations on English Bodies," p. 139.

[22] Wilson, p. 139.

[23] On the Chamberlens, see Helen King, *ODNB*, online ed., Oxford University Press, September 2004 (http://www.oxforddnb.com/view/article15062).

[24] C.H. McIlwain, ed., The *Political Works of James I* (Cambridge: Harvard University Press, 1918), p. 24.

[25] *Common Journals* (London, 1803), I (1547–1628): 184. See McEachern for discussion of Perceval and participants in the debate over the naming of the united England and Scotland, pp. 140–5.

justification—James represents his sovereignty as a nursing kingship. He addresses and represses the milk-memory of the idealized Elizabethan island-kingdom by substituting the greater breast of a masculine maternal Britain for the lesser one of a simply feminine Mother England. By offering his subjects his own all-encompassing fount of social nourishment, James not only rewrites the attachment of Scotland to England as a bond of enduring mutual affection and never-ending sustenance, but he also, through the identity-forming powers of breast-milk, unites two peoples who were perceived as morally and corporeally distinct. As Francis Bacon observes, the cultivation of love and mutual affection is crucial to the securing of Union between Scotland and England: "a union of love," he writes, is needed "to imprint and inculcate with the hearts and heads of the people that they are one People and Nation."[26]

James's nurturing, maternal aspects project Union by establishing the king as the sole source of his subjects' social nourishment, and so as the natural origin of a greater British identity. But, while the androgynous king embodies an all-encompassing oneness, his double gender also raises the specter of division and doubleness: the bad breast as well as the good one. By asserting androgynous absoluteness, James paradoxically risks opening up crevices in his own royal person and in the national surface that he endeavored to make labile. The bad breast exposed by "masculine-feminines" in *Hic Mulier* carries the very same threat of personal and national disintegration. We can better appreciate why James issued his admonition to the London clergy. The *haute couture* cross-dressed woman, with her bare breasts and sword, threatens foreclosure not only on James's self-proclaimed nursing fatherhood but also his political fantasy of an undifferentiated, seamlessly united greater Britannia, the issue of the king's sovereign paternal–maternal nurture. *Hic Mulier* finally is notable for the ways in which it both exposes and intervenes in the surprising nexus among the new ideology of maternal nurture, the reformed nation, and the expansionist fantasies of Jacobean Britain.

FROM KINGDOM TO CLINIC: OLD FATHERS AND NEW MOTHERS IN *MACBETH*

Like the female transvestite in *Hic Mulier*, Lady Macbeth is associated with the "dusky night" of the maternal matrix, to recall William Harvey's evocative phrase. The Lady calls on the spirits of darkness to "take [her] milk for

---

[26] Francis Bacon, *A Brief Discourse Touching the Happy Union of the Kingdomes of England and Scotland* (London: [R. Read], 1603), sig. C3v.

gall," and to shield her from both "compunctuous visitings of nature" (1.5.43) and the civilizing "peep" (1.5.51) of heaven. As Deborah Willis suggests, malevolent maternity first enters the play through the witches, who subvert the natural order, making fair foul and foul fair; they set the stage for the tragic action to come.[27] Unfettered neither by nature nor social convention, the hermaphrodite witches encourage men's illicit pleasures, ambitions, and feelings.[28] Most readers agree that the play relates Macbeth's insurgent, regicidal political ambitions to perverse forms of maternal nature and nurture. The witches' tantalizing prophecies feed Macbeth's desire for the throne and his willingness to assassinate the gentle King Duncan. Like the witches, Lady Macbeth, the would-be murderer of innocent infants, nurses her husband's bloody thoughts and so-called manly mettle.[29].

The much-discussed link between Macbeth's growing thirst for blood and the bleeding gums of the infant whom Lady Macbeth would rip from her breast acquires surprisingly new contours when considered in the context of the period's debate over women's worth. *Macbeth* plots out many of the same events narrated the two pamphlets. The play moves from the witches and Lady Macbeth's gall-filled breasts to Macbeth's regicide, his bloodthirsty regime and, finally, to Macbeth's death at Macduff's hand. Malcolm and his English allies redeem Scotland by demystifying the witches' prophecies when, camouflaged with leaves and branches, they literally bring "Great Birnam Wood to high Dunsinane Hill" (4.1.109)—an act that naturalizes and Anglicizes the kingdom's dark, maternal, Scottish energies. Just as the two pamphlets begin by sensationally reporting a violent eruption of unnatural motherhood (Margaret Vincent's infanticidal mania, the transvestite's sword and bare breasts) and then end by disenchanting (pathologizing or adjudicating) these disruptive implosions of the natural order, so *Macbeth* sensationally dramatizes motherhood's dark-magical energies. In Lady Macbeth, it is possible to glimpse both the maternal pitilessness of Margaret Vincent and the monstrous gender inversion

[27] Deborah Willis, *Malevolent Nurture*, p. 217. See also Dympna Callaghan, "Wicked Women in *Macbeth*: A Study of Power, Ideology, and the Production of Motherhood," *Reconsidering the Renaissance*, ed. Mario A. Di Cesare (Birmingham, NY: Medieval and Renaissance Texts and Studies, 1992), pp. 355–69.

[28] Terry Eagleton observes that the witches "inhabit an anarchic, richly ambivalent zone both in and out of official society...They are poets, prophetesses and devotees of female cult, radical separatists who scorn male power and lay bare the hollow sound and fury at its heart," in *William Shakespeare* (Oxford: Blackwell, 1986), pp. 1–3.

[29] Scholarship on the witches in *Macbeth* is immense. Two influential studies are Diane Purkiss, "*Macbeth* and the All-singing, All-dancing Plays of the Jacobean Witch-vogue," in *Shakespeare, Feminism, and Gender*, ed. Kate Chedgzoy (New York: Palgrave, 2001), pp. 216–34, and Stephen Greenblatt, "Shakespeare and the Exorcists," in *Shakespearean Negotiations: The Circulation of Social Energy in Renaissance England* (Oxford: Clarendon Press, 1988), pp. 94–128.

of *Hic Mulier*'s female transvestites. In the end, the play, like the pamphlets, demystifies and naturalizes the overblown maternal spectacle that it presents at the beginning. Thus, just as the pamphlets establish the bad mother both as a generalized social and national type and as an agent of irrational and foreign (Catholic, primitive, and Oriental) powers and social illness, so *Macbeth* focuses on Lady Macbeth's categorically bad female traits, her dangerous, alien maternal energies, and their terrible impact on the health of Scotland's body politic.

As just noted, *Macbeth*, like the pamphlets, disenchants the dark mysteries of unnatural motherhood. The play, however, takes two additional steps. First, it benignly recuperates the magic of motherhood and reassigns it to the life-giving and nurturing figure of the English king, Edward the Confessor. *Macbeth* restores magic and mystery to the traditional patriarchal figure of the monarchal sovereign. Second, in a seemingly oppositional move, it simultaneously makes room for the new "female" (or, emotionally engaged, patriotic, and disputatious) national subject to appear on stage. After learning that Macbeth's henchmen have murdered his family, Macduff, the exemplary monarchical subject, breaks with rank and tradition when he challenges Malcolm's exhortation to "Dispute it like a man" (4.3.221). Macduff anticustomarily insists that he instead must "feel it as a man" (4.3.223). Thus, rather than oppose old and new, male and female, *Macbeth*'s multi-dimensional treatment of maternal nurture mixes tradition with reform, magic with demystification, dynastic government with the new political subject.

In particular, the three consecutive scenes—Act Four, Scene Two; Act Four, Scene Three; and Act Five, Scene One—benefit from comparison with *A pitilesse Mother* and *Hic Mulier*. Act Four, Scene Three shifts the action from Scotland to England, where Macduff and Malcolm join forces against the tyrant Macbeth. We witness an exclusively male contest for power and justice. The two short scenes, Act Four, Scene Two and Act Five, Scene One, invite us into the interior, domestic spaces of Macduff's castle in Fife and Macbeth's Dunsinane. The two domestic scenes sensationalize perverse and unnatural sights: the brutal murders of Lady Macduff and her son and Lady Macbeth's mad acting-out of her unacknowledged guilt. The once-demonic Queen, Lady Macbeth, as observed by the Doctor and the Waiting-Gentlewoman, becomes the disenchanted subject of medical inquiry. Her "unnatural" (5.1.61) sleepwalking and obsessive hand-washing ("Out, damned spot" [5.1.30]) are treated, at least at first, as symptoms of disease rather than signs of witchcraft. The emphasis on the new science over the old demonology also significantly changes the social relations among the characters. Whereas, under the old order, the Waiting-Gentlewoman had hitherto attended the Lady, she now serves the Doctor, filling him in on the background of the case:

"I have seen her rise from her bed, throw her nightgown upon her, unlock her closet, take forth paper, fold it, writ upon't, read it, afterwards seal it, and again return to bed, yet all this while in a most fast sleep" (5.1.4–7). As the Waiting-Gentlewoman's detailed, eye-witness testimony makes clear, the Lady unconsciously attempts to purge her tainted character not only by her compulsive hand-washing, but also through her obsessive letter-writing. The bloody spot that the Lady imagines that she sees on her hand is strangely related to the marks that she makes in her own hand in the confession letter she repeatedly attempts to write. At the beginning of the play, the private letter serves as the powerful, secret medium through which Macbeth first informs his wife of the witches' prophecy. This letter activates the Lady's hunger for power, and her desire to be "unsexed" (1.5.). But at the end, the private letter simply becomes part of the medical record: its illicit secrets are exposed and converted into diagnosable symptoms: "I will set down what comes from her to satisfy my remembrance the more strongly" (5.1.28–9), asserts the Doctor.

At the precise moment when the new science would seem to triumph over the old magic, and when law and medicine would demystify the play's spectacular display of demonic female power, Shakespeare refuses to allow the new conceptual order to win the day. Despite his diagnostic powers, in the end, the Doctor is forced to acknowledge that the Lady's "disease is beyond my practice" (5.1.49): "she has…amazed my sight" (5.1.68). Hence, if the play begins by affiliating Lady Macbeth's perverse maternal nature and nurture with witchcraft in her unsexing speech— "Come to my woman's breasts/And take my milk for gall, you murdering ministers" (1.5.45–46)—it ends by subjecting the Lady instead to the amazed medical gaze of the pathologist, who finally is unable either to diagnose or treat her: "More needs she the divine than the physician" (5.1.65). Not unlike the two pamphlets, *Macbeth* first sensationalizes maternal deviance and then proceeds to disenchant the frightening spectacle of diabolical maternity that it creates, dismissing the witches and pathologizing Lady Macbeth. The haunted specter or "walking shadow" (5.5.23) to which Shakespeare reduces the Lady by the end empties out the wild, sensationalized presence that she assumes at the beginning of the play. As Janet Adelman writes, "By the end, she is so diminished a character that we scarcely trouble to ask ourselves whether the report of her suicide is accurate or not."[30] But, if she loses her presence and power as a diabolical Queen on the dramatic stage, she nevertheless retains the power to fascinate and amaze her audience as the female object of medical curiosity in the theater of the clinic into which the interior of Dunsinane castle is

---

[30] Janet Adelman, *Suffocating Mothers*, p. 145.

transformed.[31] If at first the Doctor dispassionately looks with scientific interest at the Lady as a medical spectacle, in the end, he is devastated by her eerie and ultimately unwatchable sleepwalking and talking: "My mind she has mated" (5.1.68). Not unlike William Harvey in his fruitless efforts to understand the mysteries of the female reproductive matrix, Shakespeare's Doctor cannot come to terms with Lady Macbeth's "infected min[d]" (5.1. 62): she marks the limit of his science. Although she retains none of her menacing malevolence, the Lady nevertheless still possesses a dark mysteriousness powerful enough to resist the radically disenchanting effects of medical observation and discourse. The Doctor's science does not protect him against the unspeakable horror he conceives after observing Lady Macbeth and her "spot": "So good night./My mind she has mated, and amazed my sight/I think, but dare not speak." (5.1.67–9). To be sure, as Adelman argues, *Macbeth* succeeds in "eliminating the female" by driving all of the female characters out of the play and celebrating Macduff as "none of woman born" (4.1.96).[32] *Macbeth*, in the end, disarms and detoxifies the Lady's unnatural, maternal bloodthirstiness. At the same time, however, as the Doctor's terror suggests, *Macbeth* prevents the new science from fully disenchanting the mysteriousness of motherhood. Whereas the two pamphlets demystify motherhood through law or medicine and translate the maternal into a subject of rational, public, and national debate, *Macbeth* keeps the magical, arcane nature of maternity at least partially intact, thereby preventing motherhood from becoming the subject of common discourse or scientific inquiry.

The savage assault on Lady Macduff and her children also presents motherhood in ways that both overlap with and differ from the two pamphlets. Natural motherhood and the public implications of the private household are the key issues here, as they are in the pamphlets. As David Norbrook observes, "Shakespeare invents the scene of the murder of Lady Macduff and her son in order to bring home the 'natural' links between the public and the private."[33] As many readers have observed, Lady

---

[31] "When I leave his clinic," Freud writes of Charcot and his female hysterics, "my mind is sated as after an evening in the theater," Ernest Jones, *The Life and Work of Sigmund Freud* (Harmondsworth: Penguin Books, 1964), pp. 173–4; Shakespeare's Doctor has a much less enjoyable "evening in the theater" than did Freud at Charcot's clinic.

[32] Adelman, p. 145.

[33] David Norbrook, "*Macbeth* and the Politics of Historiography," in *Politics of Discourse: The Literature and History of Seventeenth-Century England*, ed. Kevin Sharpe and Steven N. Zwicker (Berkeley: University of California Press, 1987), p. 104. By contrast, Robert Watson argues that *Macbeth* associates Nature with mortality and flux rather than with the autochthonous character of established social institutions such as the "natural" family and kingship in "Another Day, 'Another Golgotha'," *The Rest is Silence: Death as Annihilation in the English Renaissance* (Berkeley: University of California Press, 1994), pp. 134–5.

Macduff, the natural mother, identifies Macduff with Macbeth, when she names her husband a "traitor" (4.2.45), a term that ties Macduff to Macbeth, who from the very start of the play (when he assumes the treacherous Thane of Cawdor's title—"that most disloyal traitor" [1.3.52]) is associated with "treasons capital" (1.3.113). Scotland proves incapable of properly balancing the dynastic equation between family and state, made in the name of natural motherhood. Lady Macduff and her son die terrible deaths at the hands of Macbeth's henchmen, and Scotland succumbs to the tyrant Macbeth and to the wild and unnatural maternal energies that inspire his blood-thirst. As Adelman argues, *Macbeth* attempts to "obscure the operations of male power, disguising them as a natural [maternal] force."[34]

Like *Macbeth,* the two pamphlets celebrate nature and the natural mother, but they do so as *alternatives* to the very same traditional, blood-and-soil, forms of dynastic power that Shakespeare deems natural. For the *Hic Mulier*-author, art not nature is associated with customary land-and-birthright measures of identity, at least for women. The artful transvestite refuses to relinquish the class markers that preserve her difference from other women. By contrast, the natural mother is praised for her generic likeness to all other good women—and as the symbol of the larger oneness of national community as well. As we have seen, women cross-dressers resist the notion that "we all be coheirs of one honor, one estate, and one habit" (271)—the oneness exemplified by good women in their unadorned naturalness. The natural mother in *Hic Mulier* is rewarded for her naked virtues, especially the honest care of her children, with the highest expressions of undying love, with which she binds men's souls into a new spiritual community purer than that of the old corrupt body politic, with its emphasis on social degree and other external markers of identity: "you shall draw men's souls unto you with that severe, devout, and holy adoration, that you shall never want praise, never love, never reverence" (272).

In *Macbeth*, in contrast, natural motherhood evokes the traditional, dynastic equation between family and state. Macbeth is punished for his unnatural crimes against the dynastic line by being deprived of the very same natural fertility that organically perpetuates the dynastic line. In the end, Macbeth takes over the Lady's role as an unnatural, non-nurturing mother, who, like Margaret Vincent, tyrannizes and kills her children/subjects. As Norbrook observes, Banquo, to emphasize Macbeth's barren monarchy, points to his long line of royal offspring in the spectacle of kings that the witches conjure up for Macbeth. At the beginning of the play, Banquo sees Dunsinane castle as a teeming womb and source of

---

[34] Janet Adelman, *Suffocating Mothers*, p. 145.

overflowing nurture, best characterized by "the pendant bed and procreant cradle" of the birds that nest in its walls (1.6.8). In contrast, the non-maternal Lady within those same walls urges her husband to defy the fruitful natural-dynastic order. Lady Macduff's brutal murder further sensationalizes Macbeth's larger violence against Mother Nature—violence that in the end turns against him by denying him the power to naturally perpetuate the dynastic family/state.[35] Instead of a promising and pro-creative national future, Macbeth perceives only a tedious, sterile series of meaningless reiterations: tomorrow and tomorrow and tomorrow.

With the shift from Scotland to England, however, the traditional nation finds new vitality and significance. The good motherhood exemplified by Lady Macduff—and seemingly extinguished with her death—is revived in the person of Edward the Confessor. Malcolm marvels at the pious English king's life-giving and healing powers.[36] Edward represents the benign alternative to the witches' and Lady Macbeth's unnatural and deadly nurture. He strangely resembles the witches: like them, he has the "gift of prophecy," but his is "heavenly" while theirs is hellish (4.3.158). Not unlike the Lady's bloody hand, Edward's miraculous, healing hand exposes the limitations of medical science. The Doctor's sole purpose in the scene is to confirm that the magic of the king's touch far surpasses even the very best remedy that medicine can offer:

> There are a crew of wretched souls
> That stay his cure. Their malady convinces
> The great essay of art, but at his touch,
> Such sanctity hath Heaven given his hand,
> They presently amend.
>
> (4.3.143–5)

Medicine can neither heal the Lady's malevolent maternal nurture nor replicate the king's life-giving maternal magic, which alone "cures" and "amend[s]" (4.3.153, 146). In the two pamphlets, the natural is the space of disenchanted experience, a world organized by law, accessible to

---

[35] Norbrook, "*Macbeth* and the Politics of Historiography," pp. 104–5.

[36] One notable point of historical context for this scene is that, after his English accession, Scottish king James VI was reluctant to exercise the royal touch. When he received his first request for the healing touch at Woodstock in 1603, he repudiated its superstitious assumptions. In "The English Accession of James VI: 'National' Identity, Gender and the Personal Monarchy of England," *The English Historical Review*, 117. 472 (2002): 525, Judith M. Richards argues that James's disdain for practicing the royal touch exacerbated English distaste for their new king not only as a foreigner, but also as inferior in gendered terms to Elizabeth I, whose female sovereignty was revered long after her death. The Queen had made contact with subjects "from across the range of the social spectrum" by regularly performing the royal touch and other public rituals celebrating the sacredness of the monarchy.

empirical analysis and rational government—this is the realm of the new mother. The magical king's curative-and-creative touch benignly re-enchants Nature and the natural mother. His nurturing hand also piously re-mystifies the traditional natural equation between family and nation. Although Shakespeare reduces the once-menacing Lady to a disenchanted set of pathological symptoms, he also leaves the core mysteriousness of motherhood intact, so that it can be converted to good maternal use by the English king, a God-like savior and nurturing redeemer. Whereas the bad, unnatural mother is opposed to the good, natural mother in the two pamphlets, in *Macbeth*, the bad mother not only is driven out of the play, but her bad nurture also is redeemed, re-mystified, and reassigned to the English king. Edward appropriates motherhood's rationally impenetrable mysteries but not its malevolent unruliness, which is purged from the play along with the witches and Lady Macbeth.

Although a bit removed from the immediate context of *Macbeth*, Joad Raymond's observation that the "*exposure* of the [Charles I's] person...was one of the great ideological tremors of the 1640s" nevertheless helps to further illuminate the implications of Shakespeare's re-enchantment of English kingship in 4.3.[37] For all its spectacular blood and horror, *Macbeth* creates no notable "ideological tremors" in relation to kingship. Rather the play shrouds its royal male characters in maternal mysteries, which as the Doctor in both 4.3 and 5.1 observes, cannot be rationally or publicly exposed, either visually or discursively. Rather than disappear with the witches and Lady Macbeth, these unfathomable maternal mysteries resurface in sacramental political form in both the magic of monarchy and in the rationally irreducible, androgynous absoluteness of kingship. Like Edward the Confessor, but less perfectly, Duncan, as an androgyne, embodies the sacred mysteries of kingship in maternal terms. Of Duncan's feminized meekness, Macbeth reflects:

> Duncan
> Hath borne his faculties so meek, hath been
> So clear in his great office, his virtues
> Will plead like angels, trumpet tongu'd against
> The deep damnation of his taking off...
> (1.7.16–20)

This depiction of Duncan tellingly departs from its source in Holinshed's *Chronicles*, where the gentle king is an object of deep scorn, a "faint-heart

---

[37] Joad Raymond, "Popular Representations of Charles I," in *The Royal Image: Representations of Charles I*, ed. Thomas N. Corns (Cambridge: Cambridge University Press, 1999), p. 47.

milke sop," whose overly compassionate nature facilitates the revolt of Makdonwald, which Macbeth is sent to put down.[38] Unlike his source, Shakespeare never scorns the "milke-sop"; instead, he suggests that Duncan derives his "great office" (1.7.18) or his sacred kingly authority from his milk and milky nature, which "will plead like angels" (1.7.19). In the same way, Edward the Confessor, although peripheral in every way to the play's action and never actually present on stage, nonetheless serves as the play's central life force. Of the highest divinity, his miraculous curing powers, the king's touch, are also the stuff of the very same childish superstitions that Lady Macbeth skeptically mocks at the beginning of the play. In her clinical, iconoclastic view of death, the murder of the king cannot create supernatural tremors, vengeful ghosts, or divine retribution: "The sleeping and the dead/," she maintains, "Are but as pictures. /'Tis the eye of childhood/That fears a painted devil" (2.2.51–53). But, as her brutal fantasy of killing the babe at her breast underscores, the Lady's cynical perception of "the eye of childhood" is not to be trusted. At the end of the play, the Lady's clinical view of the sleeping and dead is turned against her, when, in the sleepwalking scene, the Doctor reduces her to a medical specimen—but, as already noted, just enough to reduce her to a set of female symptoms, yet not enough to wholly master her maternal magic, which is appropriated, redeemed, and Anglicized by the saintly King Edward. Charles I's unwillingness to touch for the King's Evil and to drape himself in the mysterious maternal garments of royalty helped to facilitate not only Charles's exposure, but also the desacralizing and decentering of the Stuart monarchy during the civil wars. Whereas in the two pamphlets science and law triumph over superstition and magic, in *Macbeth* magic both benefits from and subsumes science and reason.

Just as the English king redeems the malevolent maternal magic of the witches and Lady Macbeth, so, the play implies, shall England redeem Scotland, the bad, unnatural mother country. Shakespeare approaches the question of Union with characteristic diplomacy: he never spells out the precise relationship between the two kingdoms. Although Malcolm pays homage to Edward, Edward is not brought on stage and so cannot upstage the legitimate Scottish heir. Malcolm accepts aid from England, but he and Scotland are never put into vassalage by the English king. Indeed, Malcolm's future relations and obligations to Edward are never articulated. The play nevertheless points to Anglo-Scottish Union, even while rendering that federation in open-ended terms. The English king's appropriation and purification of the witches' and the Lady's maternal

---

[38] Raphael Holinshed, *Chronicles*, ed. E.G. Boswell-Stone (London: Lawrence, 1986), p. 9.

magic allow for a projection of the same kind of uncontested and unconditional display of kingship's androgynous absoluteness and British
Union that James, less successfully, cultivates in his self-image in *Basilikon Doron* as a nursing king and in his 1620 admonition to the London
clergy against female cross-dressing.

But while *Macbeth* sanctifies absolute kingship, it also makes room,
however briefly, for the emergence of the new emotionally and intellectually engaged political subject, fostered by the new print culture. As in the
two pamphlets, this new subject position emerges through a debate about
the two sexes, when Malcolm and Macduff argue about value of feminine
feeling after Ross tells Macduff about the brutal murder of his family.[39] In
response to Macduff's overwhelming grief, Malcolm urges, "Dispute it
like a man" (4.3.221), to which Macduff counters, "I shall do so/But
I must also feel it as a man" (4.3.222–23). This debate revisits the debate
about gender values between the Macbeths at the beginning of the play.
Not unlike Lady Macbeth, Malcolm construes manhood as the stopping
up of feminine feeling and milky compassion. Macduff insists that to be
a man he "must...feel" the magnitude of his wife and son's deaths and his
own unwitting complicity in them as if in answer to his wife's accusation
in 4.2.9 that "He loves us not,/He wants the *natural* touch" (emphasis
added).

Here as elsewhere in the play, the nature of Nature and maternal
nature prove to be unsettled and unsettling concerns. Malcolm would
have Macduff balance his private feelings and public duties so as to restore the natural link between family and the old nation, which Macbeth's crimes against the dynastic order brutally destroy. Following heroic
convention, Macduff at first had placed the public over the private, but
he later allows the private to overwhelm the public. Macduff's active resistance to Malcolm's demand that he balance the natural equation between these two spheres in the interest of dynastic government results in
a new form of public–private imbalance. Even more importantly,
Macduff's passionate display of feeling also opens the new space necessary for an alternative vision of nature and nation to surface in the play.
Rather than serve his king, Macduff wants to act independently on his
own grief-stricken expression of love for his wife—to place his tender
feelings for the natural mother of his children in service of something
different from the old organic model of manhood and nationhood that

---

[39] Sharon Achinstein comments on the generic connections between the debate and the
drama in *Milton and the Revolutionary Reader* (Princeton: Princeton University Press,
1994), pp. 103, 110. See also Margot Heinemann, *Puritanism and Theatre: Thomas Middleton and Opposition Drama under the Early Stuarts* (Cambridge: Cambridge University
Press, 1980).

inspires Malcolm's campaign to reclaim the throne from Macbeth and repair the dynastic line.[40]

Macduff's vague but resistant desire for "something else" never coalesces into anything concrete. When, however, read in relation to the *Hic Mulier*-author's description of the binding spiritual power that the good mother has over men's souls ("you shall draw men's souls unto you with that severe, devout, and holy adoration, that you shall never want praise, never love, never reverence" [272]), the subversive political potential of Macduff's overwhelming outpouring of emotion for the idealized figure of his dead wife and natural mother of his children becomes more transparent. In *Hic Mulier*, the good mother's consolidating social energies refine and almost abstract the old body politic out of existence. Like *Hic Mulier*, *Macbeth* implicitly links the potential for political opposition to the dynastic nation with passionate feeling for the mother. In *The True Lawe of Free Monarchies*, James I repudiates those who justify political rebellion by appealing to men's "naturall zeale" for the mother: "It is casten vp by diuers, that employ their pennes vpon Apologies for rebellions and treasons, that euerie man is borne to carrie such a naturall zeale and duety to his common wealth, as to his mother" (66). If, on the one hand, after Lady Macduff perishes, her natural maternity is recuperated and re-enchanted by the mysterious figure of the life-giving English king, on the other, Macduff's feminized expression of "naturall zeale" for his dead wife, who is the exemplar of the natural, nurturing mother, implies rebellion against the dynastic nation. Macduff ultimately complies with the traditional, dynastic measures of manhood and nationhood, which are ratified by Malcolm: "This goes manly" (4.3.237). *Macbeth* nevertheless briefly creates space for the new disputatious and upstart national subject to emerge on stage.

The importance of Macduff's brief but passionate protest against the old patriarchal nation is perhaps best gauged by the dispute about the manly feeling that breaks out once again in the very last scene of the play. In the short exchange between Malcolm and old Siward about how properly to mourn for the dead, Malcolm assumes Macduff's reformed stance

---

[40] Macduff comes close to "the citizen" and "the Protestant," as characterized by Helgerson: "figures who identify strongly with the nation and its ruler but both of whom are intent on keeping some part of themselves and their community free from the encroachment of national power." For Helgerson, however, "Neither of these figures is of much interest to Shakespeare." See *Forms of Nationhood: The Elizabethan Writing of England* (Chicago: University of Chicago Press, 1992), p. 197. In *Citizen-Shakespeare: Freemen and Aliens in the Language of the Plays* (New York: Palgrave Macmillan, 2005), John Michael Archer demonstrates that the language of Shakespeare's plays, especially the city comedies and Roman tragedies, conveys "[t]he tension between exclusion and inclusion in citizenship, indeterminacy, struggle, and even ethnic antagonism," p. 5.

towards "feel[ing] it as a man." In response to old Siward's conventionally stoic, soldierly response to his young son's death on the battlefield, Malcolm protests: "He's worth more sorrow,/And that I'll spend for him" (5.11.17–18). Malcolm recognizes the symbolic political capital suggested by Macduff's tender sentiments and is willing to spend it. Old Siward, however, has the last word: "He's worth no more" (5.11.19). Whether or not Malcolm's "more" (5. 11.17) ultimately will prevail over Siward's "no more" (5.11.19) remains unclear.

Old and new, male and female, magic and science, monarchical absolutism and modern national sensibility, thus all come into incoherent play in *Macbeth*, reflecting the continuity and collusion between these seemingly opposed terms. Through the trope of the maternal breast, the play conveys the cultural ambivalence and boundary panic that characterize this fraught historical moment. The maternal valences of these unresolved tensions make it possible for James I in *Basilikon Doron*, not unlike Malcolm in the last scene of *Macbeth*, to politically exploit the reformed language of mother-love and nurture, even while at the same time solidifying traditional measures of dynastic kingship. As we shall see next, like *Macbeth*, *Basilikon Doron* exposes and exploits the complex and precarious equilibrium between the old and new sides of the early modern cultural-and-gender divide.

3

Nursing Fathers and National Identity

James I, Charles I, Cromwell, and Milton

In *Basilikon Doron*, James I describes himself as a "nourish father" who succors his subjects with "nourish milke."[1] In the winter of 1651–52, the annalist of the Independent Church founded in Cockermouth praised Oliver Cromwell as the "nursing father of the churches," a description that resonates with Cromwell's own tender imagery of maternal succor and unconditional love in his parliamentary speeches and letters.[2] "Therefore I beseech you," Cromwell tells the Barebones Parliament, "…have a care of the whole flock! Love the sheep, live the lambs; love all, tender all."[3] Not unlike Cromwell, as we have seen, in Puritan New England, ministers represented themselves as "breasts of God" and their congregations as "New born babes desiring the milk of the Word" (Lev. 146). "Ministers are your Mothers too," writes Cotton Mather in a funeral sermon for his father, "Have they not Travailed in Birth for you that a CHRIST may be seen formed in you? Are not their Lips the Breasts thro' which the sincere Milk of the Word has passid unto you, for your Nourishment."[4] In "Longing," George Herbert parallels the relationship between the mother and the suckling infant with the relationship between God and the penitent sinner:

> Mothers are kind, because thou art,
> And dost dispose
> To them a part:
> Their infants, them; and they suck thee
> More free.[5]
>
> (14–16)

---

[1] C.H. McIlwain, *The Political Works of James I* (Cambridge: Harvard University Press, 1918), p. 24. All references to James I's writings are taken from this volume and noted in the text.

[2] Quoted in Claire Cross, "The Church in England 1646–1660," in *The Interregnum: The Quest for Settlement*, ed. G.E. Aylmer (Hamden, CT: Archon-Shoe String, 1972), p. 117.

[3] *The Writings and Speeches of Oliver Cromwell*, ed. Wilbur Cortez Abbott (Cambridge: Harvard University Press, 1937), III. 62.

[4] Cotton Mather, *A Father Departing…* (Boston, 1723), pp. 22–3.

[5] George Herbert, *The Works of George Herbert*, ed. F.E. Hutchinson (Oxford: Clarendon Press, 1941).

The musician Thomas Ravenscroft remembers his teachers at Gresham College, London as "kinde nursing fathers."[6]

As these examples suggest, the figure of the nursing father circulated in early modern England through a range of genres: political speeches, sermons, letters, and religious verse, among others. In this chapter, I am primarily interested in political applications of this trope. I argue that the figure of the nursing father emblematizes reformed male political authority, placing it in sharp contrast with representations of the harshness and cruelty of the old Roman Catholic patriarchy. As we shall see, the nursing father also is a highly contested figure, one that is adopted for very different political objectives in both the innovative and traditionalist Protestant writings that this chapter investigates: James I's *Basilikon Doron* (1603), Charles I's *Eikon Basilike* (1649), Cromwell's parliamentary speeches of the 1640s and 1650s, and Milton's *Of Education* and *Areopagitica*, both published in 1644. With the exception of *Basilikon Doron*, all of the texts analyzed in this chapter date to the 1640s and 1650s. Rather than discuss them in strict chronological order, however, I approach them by theme: Stuart and anti-Stuart. The first half of my chapter focuses on James I and Charles I; the second half focuses on Cromwell and Milton. This thematic grouping, I believe, best exposes the shifting political fault lines in the trope of the nursing father in the 1640s and 1650s.

Early modern figurations of the nursing father with their incongruous associations of maternal plenitude with paternal potency find their primary antecedents in scripture. As already noted, Protestants allegorized the passionate expression of erotic desire in the Song of Songs as the spiritual love of the Church for its penitent sinners. This allegorical reading of the Song inspires the striking preponderance of breasts, wombs, and other kinds of maternal imagery in the sermons of Thomas Hooker, Cotton Mather, and other Puritan divines who present themselves as the breasts of God; they suckle their newborn congregants and fill them with the milk of the divine Word. Drawing on Canticles, John Cotton writes that ministers should "apply themselves to the estate of their people: If they bee babes in Christ, to be as breasts of Milke to suckle them."[7] In addition to the Song, the nursing father evokes two other resonant Hebrew scriptural passages: Isaiah 49:23: "And kings shall be thy nursing fathers," and Numbers 11:12, where Moses, somewhat petulantly, describes himself as a reluctant leader who must carry

---

[6] Quoted in Linda Phyllis Austern, "'My Mother Musicke': Music and Early Modern Fantasies of Embodiment," in *Maternal Measures*, p. 249.

[7] John Cotton, *A Briefe Exposition of the whole Book of Canticles...* (London, 1642), pp. 240–1.

his people in "[his] bosom, [as] a nursing father beareth a sucking child."
Evoking Isaiah 49:23: "And kings shall be thy nursing fathers," James in
Basilikon Doron depicts himself as a "louing nourish father,"—as father
and mother to his people and so absolute in sexual terms. All-encompassing, the king is above comparison. "Remember," he writes to his son
in *Basilikon Doron*, "that...[God] hath erected you aboue others. A
moate in anothers eye, is a beame in yours" (12). Cromwell and his
apologists also deploy the figure of the nursing father, but they translate
it into an emblem of non-monarchical political power. Evoking Moses
as the nurturing but reluctant leader of Numbers 11:12, Cromwell represents himself as an inspired emancipator—and not a self-aggrandizing
conqueror. Like his Hebrew prototype, he delivers his oppressed people
from bondage (from the Norman yoke) and, with God's help, he nourishes them with the manna of liberty.[8]

The trope of the nursing father also was familiar to early modern
Christians through New Testament figurations of the deity. The image
of Jesus as mother finds especially vivid expression in the writings of
Julian of Norwich: Jesus is "a kind nurse that hath naught else to do
but to attend to the well-being of her child," who feeds us from "his
blessed breast," and to whom we can say: "my most dear Mother, have
mercy on me. I have made myself foul and unlike to thee; and I cannot
or may not amend it but with thine help and grace."[9] As just noted,
this venerable medieval image of Jesus as nurse resurfaces in George
Herbert's poetry, but with a new twist. Herbert celebrates and, at times,
conflates his mother, Magdalen Herbert, with the nursing Jesus. As
Robert Watson observes, Herbert's Christianity is a "maternally imbued
faith...[he] sees his mother through a glass darkly, and recognizes his
deity."[10] As many readers have noted, in *Eikon Basilike*, Charles I assumes the role of a Christ-like martyr. As we shall see, however, the
king also fashions himself in the affecting image of Jesus as a loving
nurse. He does so for two main reasons: to compensate for his diminished patriarchal power by adopting a potent, divine image of feminized
masculinity; and to cultivate his people's affective attachment to him as
their sole source of spiritual sustenance, upon which they could feed

---

[8] Invoked time and again, in radical and republican texts, the Exodus story provides a
narrative framework and biblical justification for seventeenth-century ideas of liberty, revolution, regicide, and the new Israel. See Michael Walzer, *Exodus and Revolution* (New York:
Basic Books, 1985).

[9] Julian of Norwich, *Revelations of Divine Law*, trans. James Walsh (New York: Harper,
1961), pp. 164–8. Carolyn Bynum, *Holy Feast and Holy Fast: The Religious Significance of
Food to Medieval Women* (Berkeley: University of California Press, 1987).

[10] Robert Watson, *The Rest is Silence: Death as Annihilation in the English Renaissance*
(Berkeley: University of California Press, 1994), p. 259.

even after his death. We shall return to these issues in a later section of this chapter.

In addition to their scriptural resonances, early modern figurations of the nursing father reflect the period's enhanced secular preoccupations with maternal breast-feeding as a natural means to forge a strong English identity and to create unbreakable bonds of national affection. As discussed in Chapter 2, James I uses the trope of the nursing father to strengthen his arguments for Union. But, the trope also permits James to articulate other kinds of expansionist desires under the cover of paternal–maternal love and nurture. By depicting himself as a "louing nourish father," James hopes to conquer the hearts of his subjects—to attach their thoughts and emotions to his nursing breast. Like Shakespeare's Henry V, the king wants to do more than command his subject's knee; he also wants to gain control over the "health of it" (*Henry V*, 4.2. 239) by fostering the people's willing devotion. As a nursing father, James appropriates the nourishment, love, and charity attributed to the new nursing mother and extends them to his subjects.

It is true that in *The True Lawe*, James contemptuously associates the rhetoric of mother-love with those who "employ their pennes vpon Apologies for rebellions and treason": "It is casten up by divers, that employ their pennes vpon Apologies for rebellions and treasons, that euery man is borne to carrie such a naturall zeale and dutie to his commonwealth, as to his mother" (66). By deploying the word "zeale," a key term of anti-Puritan invective, the king connects the new ideology of nurture and mother-love with radical sectarianism. Claire McEachern observes that "For James, it is not feeling for the mother which should determine civic identity, but duty to the father—not love, but law."[11] McEachern does not discuss James's self-image as a "louing nourish-father" (24). If we read *Basilikon Doron* in relation to *The True Law* (a reading warranted by the king's 1603 volume, dedicated to his English readers, in which the two tracts appear together), we see that, in addition to opposing "feeling for the mother" with "duty to the father," James also integrates maternal feeling with paternal duty. On the one hand, the king repudiates "naturall zeale" for the mother as inspiring "rebellions and treasons" (66). On the other hand, however, by depicting himself as a "nourish-father" (24), he not only encourages the same zealous expressions of maternal love, but also redirects these seditious maternal passions to his role as male sovereign, in which persona he quells the dissent that he cautions against in *The True Law*. As an

---

[11] Claire McEachern, *The Poetics of English Nationhood,* 1590–1612 (Cambridge: Cambridge University Press, 1996), p. 22.

androgynous mother–father, James connects civil obedience and national identity to both "feeling for the mother" and "duty to the father,"
to love and *True Law*.

Like James, Charles I in *Eikon Basilike* deploys the rhetoric of mother-
love and nurture to silence dissent and expand his authority over his
people by winning their hearts. Writing after his defeat on the battlefield and with a death sentence on his head, however, Charles's politically motivated uses of the rhetoric of maternal nurture inevitably differ
from his father's. Charles cultivates love and devotion in his subjects so
that he can affectively engage their memories—to ensure that they will
remember him as he wishes to be remembered and not as the opposition
would have him be preserved for posterity: as an uncaring tyrant brought
to justice by loyal English patriots. As the multiple printings of *Eikon
Basilike* suggest, Charles proves the victor in the battle over cultural
memory. Although he loses his life and crown to his enemies, he ultimately wins the war for hearts and minds. Through his text's many displays of his love and selfless devotion to God, his children, and his
people, and through the pretense of revealing his heart and nurturing
character to the readers of *Eikon Basilike*, the king paints himself for
posterity as a Christ-like nurturer. The enhanced symbolic value and
significance Charles gains by defining himself in the image of Jesus as
nurse compensate for the concrete losses he suffers as a result of collapsed patriarchal authority.

Cromwell's public speeches similarly exploit the emotional richness
and political utility of the language of maternal love and nurture. Not
unlike James and Charles, by speaking of maternal tenderness and love,
Cromwell as a nursing father seeks to win the hearts and minds of the
English people—a people stirred by the desire for Reformation, but
shaken by regicide and civil war. Whereas the language of maternal love
and nurture lends a tender veneer to James's aggressive desires for imperial
expansion, and whereas it helps Charles to inspire his subjects' sympathies
and turn them against his political enemies, the same nurturing rhetoric
enables Cromwell to convey a deeply felt sense of union to a nation radically uprooted by civil war and revolution—a union forged at the breast
of the nursing father. In his early speeches, Cromwell expediently deploys
the language of maternal love and nurture to inspire a passionate, collective imagining of the national whole that could substitute for the politically fragmented English state in the aftermath of civil war.[12] Asserting
ancient and modern, sacred and secular authority, Cromwell as nursing

---

[12] Benedict Anderson, *Imagined Communities: Reflections on the Origin and Spread of
Nationalism* (1983; rpt. London: Verso, 1991).

father addresses the period's heightened anxieties about death, continuity, and origin by campaigning for his people's heartfelt commitment to an abstract vision of a unified, post-dynastic England. This vision proved difficult to implement in institutional terms. The gap between national ideal and institutional practice widened during the Protectorate years. Cromwell's language of maternal love and nurture becomes increasingly unstable and equivocal: it both evokes and undercuts Stuart configurations of the English monarch as a nursing father. By speaking of love and maternal tenderness, Cromwell as Lord Protector simultaneously addresses the need for religious toleration, and he expands his personal rule and king-like influence over the English people. He commands obedience, and he fosters liberty of conscience.

In contrast to Cromwell's equivocal Protectoral speeches, the language of mother-love and nurture in Milton's *Of Education* and *Areopagitica*, both published in 1644, strongly argues the need for a new gentle, charitable form of Protestant male authority. In both tracts, Milton maintains that tyrannical fathers, preceptors, magistrates, and kings produce frightened and slavish children, students, and political subjects. Rather than force consciences, the reformed family/nation and school would gently nurture their charges into independent thinkers and activist citizen-patriots, ready and willing to serve the cause of national reformation. Milton's tracts explicitly demonstrate the close interplay between familial and civic reform in English republican discourse.[13] Although he avoids imagery that concretely conjures up the maternal breast, he nevertheless evokes the abstract virtues, such as charity, long associated with the nursing mother. He maintains that a truly charitable government fosters free, independent-minded, and mature citizens, while traditional patriarchal harshness and severity, as exemplified by the Spanish Inquisition, stall the pace of Reformation by stunting the English people's intellectual and emotional development, which (as in the guidebooks examined earlier) should begin at the breast. The rhetoric of nurture in *Areopagitica* additionally highlights Milton's profoundly anxious relationship to the Hebraic precedent—especially, to Moses as a nursing father. It reveals politically disabling tensions in the poet's understanding of relations between the Hebraic and the Christian, the paternal and the maternal, law

---

[13] Norbrook adeptly summarizes the complex divisions and confluence between private and public in republican thought: On the one hand, "a general preoccupation of republican thought was the need to reclaim a public sphere from the private monopolies of church and state." On the other, "Puritan ideology laid special emphasis on the household as an instrument of godly reformation which could do its work even when the public world was corrupt. A sharp public–private split was thus broken down: if under 'modern prudence' the state became a household, there were counter-tendencies toward making the household a *polis*." *Writing the English Republic*, pp. 116–17.

and love. In later chapters, we shall explore how Milton attempts to reconcile these Hebraic-Christian tensions through the discourse of mother-love and nurture. But first, the striking convergence between the language of maternal nurture and the self-fashioning of Stuart kingship needs more specific attention.

BASILIKON DORON AND THE "NOURISH-FATHER"

In his classic study of the rise of the modern family, Philippe Ariès argues that the late sixteenth and early seventeenth centuries witnessed a radical transformation in the institutional function of the family and the relations between parents and children. "The family," Ariès maintains, "ceased to be simply an institution for the transmission of a name and an estate, it assumed a moral and spiritual function: it molded bodies and souls. The care expended on children inspired new feelings, a new emotional attitude to which the iconography of the seventeenth century gave brilliant expression."[14] For Ariès, the late sixteenth and early seventeenth centuries represent that threshold moment when the family becomes modern. This historical shift finds especially clear expression in the period's "discovery of childhood": the new feelings that reformed parent–child relationships inspire, as can be seen in seventeenth-century images and texts about family life.

Published first in Scotland in 1598 and then republished with *The True Law of Free Monarchies* in England in 1603 when James ascended to the English throne, *Basilikon Doron* can be fruitfully read in relation to Ariès's account of the early modern family.[15] *Basilikon Doron* is written as an advice book, given as a gift from the king to his son, Prince Henry. After

---

[14] Phillipe Ariès, *Centuries of Childhood: A Social History of Family Life*, trans. Robert Baldick (New York: Random House/Vintage Books, 1962), p. 142.

[15] Excellent studies partly inspired by Ariès's landmark work include: Jay Fliegelman, *Prodigals and Pilgrims: The American Revolution Against Patriarchal Authority, 1750–1800* (Cambridge: Cambridge UP, 1982); Jonathan Goldberg, "Fatherly Authority: Politics of the Family," in *James I and the Politics of Literature* (Baltimore: Johns Hopkins University Press, 1983), pp. 85–112; Debora Shuger, "Nursing Fathers: Patriarchy as a Cultural Ideal," *Habits of Thought in the English Renaissance: Religion, Politics, and the Dominant Culture* (Berkeley: University of California Press, 1990), Chapter 6; Alan Macfarlane, *Marriage and Love in England: Modes of Reproduction, 1300–1840* (Oxford: Oxford University Press, 1986); Leah Sinanoglou Marcus, in *Childhood and Cultural Despair* (Pittsburgh: University of Pittsburgh Press, 1978). In *The Rise of the Egalitarian Family* (New York: Academic Press, 1978), Randolph Trumback sees the seventeenth century as the point of transition between patriarchalism and egalitarianism in the family. For an opposing view, see Lawrence Stone, *The Family, Sex and Marriage in England, 1500–1800* (New York: Harper & Row, 1977). Stone argues that the coercive power of parents increases from the sixteenth century onwards.

its 1603 republication, to which was added a preface to the English reader, the book enjoyed an unprecedented popularity. Nine versions of the text and a total of twelve thousand copies were printed in London in 1603 alone. While the large audience for *Basilikon Doron* reflects the nation's curiosity about its new king, the enormous public response to the text, I argue, also testifies to the period's new interest in and anxieties about the changing forms and functions of parent–child relations. In part, *Basilikon Doron* finds a wide audience because it addresses the period's divergent but equally intense preoccupations with the royal family and the reformed family. James's text generates enormous interest because the book's profound ambivalences, double audience and aims, and conflicted treatment of father–son, king–subject relations speak from both sides of the conceptual divide wrought by the emergent shift in early modern cultural perceptions of family life and child-rearing. This divide was to unsettle the customary forms and functions of the English family, and the royal family most especially.

Basilikon Doron's bifurcated perceptions of the family find particularly clear expression in the king's shifting old and new representations of father–son relations. On the one hand, with its resolute focus on filial respect for and deference to parents, *Basilikon Doron* endorses a traditional model of the institutional function of the family. Time and again, James evokes the Fifth Commandment, "Honour thy father and thy mother" (Exodus 20:12); he argues that the optimal way for a king to command the love of his subjects is to protect and enshrine the legacies of his predecessors. So crucial is it for a monarch to safeguard the sacred memory of his parents that in the second section of the text, which focuses on the king's conjoint role as both judge and law-giver, James ranks "the false and unreuerent writing or speaking of malicious men against your Parents and Predecessors" as chief among "vnpardonable crimes." Alluding once again to Exodus 20:12, he states: "Ye know the command in Gods lawe, *Honour your Father and Mother*" (21). "For how can they loue you," James asks, "that hated them whom-of ye are come?...It is therefore a thing monstrous, to see a man loue the childe, and hate the Parents: as on the other part, the infaming and making odious of the parents is the readiest way to bring the sonne in contempt." Parental will—their "blessing or curse"— determines the destiny of children; a child's deference to his parents, "your superiors," accord with both the natural order and civilized one as well. As James proclaims: "O inuert not the order of nature by iudging your superiours, chiefly in your owne particular! But assure your selfe, the blessing or curse of the Parents, hath almost euer a Propheticke power ioyned with it" (41). In terming Henry his "Natvral Successovr" (3), James celebrates the traditional, organic measures of status, entitlement, and inheritance that organize the dynastic nation. Invoking Exodus 20:12, the king cautions his son that the strength and longevity of his future sovereignty will depend upon the devotion he shows to his forbears. "And if there were no more, honour your Parents, for the lengthening of your owne dayes, as God in his Law promiseth" (11).

James reinforces the natural display of filial devotion and subjection to the assertion and perpetuation of royal power by compounding Henry's filial obligations. Not only must the son honor his own parents, but he must "Honour also them that are *in loco Parentum* vnto you, such as your governours, vpringers, and Praeceptors: be thankefull vnto them and reward them, which is your dewtie and honour" (41). The circular structure of this last sentence, which begins and ends with "Honour," highlights the equation that James makes between honor given and honor received. The paradigm of reciprocal giving and receiving evoked here and elsewhere in the tract suggests gift-exchange, as underscored by *Basilikon Doron*'s title. These are the primordial attachments and obligations that James fosters by giving his book as a gift to his son.[16] From the very beginning of James's text, the king makes the imperative for worthy return absolutely clear. Through a witty bilingual pun in the dedicatory address, James writes that his heir is "*rather borne to* onus, *then* honos" (3). Whether or not James will use his "Propheticke power" (41) as a father to bless or curse his son depends entirely on the prince's ability to meet his obligation to give his thanks in ways that match the worth of the king's gift of his book:

> I charge you, as euer yee thinke to deserue my Fatherly blessing, to follow and put in practise, as farre as lyeth in you, the praecepts hereafter following. And if yee follow the contrary course, I take the Great GOD to record, that this Booke shall one day bee a witnesse betwixt mee and you; and shall procure to bee ratified in Heauen, the curse that in that case here I giue vnto you (5).

James also describes his book as "my Testament and latter will" (4). To fulfill his obligation to his father's gift, Henry, as the executor of his father's will, not unlike Portia in *The Merchant of Venice*, must abide by the dictates of his father's book, even after the king's death: "by the presence of GOD and the fatherly authoritie I have over you, that yee keepe it euer with you." "[T]his Booke...," writes James, "I ordaine to be a resident faithful admonisher of you" (4). James's book expresses the king's "will" in every sense of that word: as a command, an expression of desire, a legal document, and as the binding obligations imposed by gift-exchange.

---

[16] Marcel Mauss, *The Gift: Forms and Functions of Exchange in Archaic Societies*, trans. Ian Cunnison (New York: Norton, 1967), pp. 40–1. Mauss observes that "the obligation to give" and "the obligation to receive" are equally "constraining," but the "obligation of worthy return is imperative."

Striking a highly affecting note, James ends his dedicatory address to Henry by bestowing his heartfelt blessing to his son: "I end, with my earnest prayer to GOD, to worke effectually into you, the fruites of that blessing, which here from my heart I bestow vpon you." The king desires that the many-reprinted copies of his book will serve "as witnesses to my Sonne, both of the honest integritie of my heart, and of my fatherly affection and naturall care towards him" (5). Just as James customarily addresses the prince's dynastic role as his "Natvral Successor," so the king also underscores his "naturall care" (5) or nurture: he thus speaks in the old organic rhetoric of dynastic kingship and in the new rhetoric of natural nurture at the same time. Through "*The honest integritie of my heart*" and his "fatherly affection" (5), James not only renews the customary organic measures of his "will," i.e. his royal sovereignty, his dynastic legacy, and the obligations of gift-exchange, but he also creates new abstract and affective social bonds between himself and his people. By depicting himself as a nursing father, the king appropriates the reformed nexus among nature, maternal nurture, and the nation and inserts these newly affiliated concepts into the traditional framework of the dynastic body politic.

These odd contiguities between the old, embodied paradigm of the family/nation and the new, disembodied one come to the surface in unexpected ways throughout James's text. James describes marriage as the best way for his son to fulfill both his customary familial obligations and his quasi-legal, contractual ones as well. He can pay back the honor he owes his parents by marrying well and by creating his own princely heir and debtor, an exact replica or copy of himself. Self-replication also allows the Prince to make an even exchange with the king: biological copy or son and natural heir for the printed copy of his father's gift-book, which contains his will and legacy. By producing a natural heir, Henry will carry out his father's will by continuing the dynastic line. At the same time, he fulfills the implied contractual obligation in gift-giving for a worthy return. He does this by sustaining the circular relation between honor given and honor received, which is associated in the text with gift-exchange. A king is both a lamp and a mirror—"your person should be a lampe and mirror to your company" (37). He illuminates the highest virtues and consolidates his authority and power by means of mirror-like, dynastic self-replication. James's book is "*that mould whereupon* [Henry] *should frame his future behaviour*" (5). Like his father, Henry must replicate himself through a son who will reflect his honor, just as he reflects the honor of his parents. By producing his own self-replicating heir, Henry completes and perpetuates the circle of gift-exchange and the quasi-legal obligations that it exacts.

Notably absent from this mirror-model of dynastic self-replication and the gift-cycle of debt and repayment, as many readers have observed, is acknowledgement of the mother—and, more specifically, the king's own highly problematic mother, Mary Queen of Scots, from whom James inherited the Scottish crown and derived his claim to the English throne.[17] The Queen's ambiguous will and legacy—her mixture of maternal blessing and curse—can be felt in the meticulous attention that James pays to managing his children's memories of their parents in *Basilikon Doron*. James's recurrent motif of filial devotion and parental honor is motivated in part by his desire to downplay not only Mary's notoriety, but also his own deeply flawed role as her son: his motivated silence at and acquiescence in his mother's beheading. James repeatedly tries to exonerate both his dead mother and himself. He also exhorts Henry to display perfect filial devotion to his mother, Queen Anne: "And if it fall out that my Wife shall out-liue me, as euer ye thinke to purchase my blessing, honour your mother: set *Beersheba* in a throne on your right hand: offend her nothing, much lesse wrong her" (41). If children are perfect reflections of their parents, then Henry's unblemished relationship to his mother will necessarily reflect, and correct, the king's own image as a son. The same desire to rewrite the past by stage-managing the future informs James's characterization of kingship as a lamp. By shining before his subjects as a beacon of virtue and "glistering" (12) reflection of divine righteousness when he ascends to the throne, Henry will perfect his father's image: "Remember then, that this glistering worldly glorie of Kings is giuen them by God to teach them to presse so to glister and shine before their people in all workes of sanctification and righteousnesse" (12). *Basilikon Doron*'s many aphorisms on honoring parents thus do a great deal of political work.

Whereas James relies on the old familial paradigm in his discussion of marriage and filial duty and absents his own mother, he nevertheless introduces aspects of the new primacy of maternal nurture in his depiction of kingship. The theater, as often has been noted, provides James with his most important metaphors.[18] His understanding of the king's role as a lamp and a mirror merges perfectly with his perception of kingship as a perpetual, exemplary performance on the public stage: a "King is as one

[17] Jonathan Goldberg discusses James I's complex relationship to his mother in *James I and the Politics of Literature*. Coppelia Kahn analyzes Jacobean occlusions of the mother in relation to *King Lear* in "The Absent Mother in *King Lear*," in *Rewriting the Renaissance: The Discourses of Sexual Difference in Early Modern Europe*, ed. Margaret W. Ferguson, Maureen Quilligan, and Nancy J. Vickers (Chicago: University of Chicago Press, 1986). For a similar argument concerning *The Tempest*, see Stephen Orgel, "Propero's Wife," in Ferguson, Quilligan, Vickers, eds. *Rewriting the Renaissance*.

[18] Stephen Greenblatt discusses Elizabeth's self-theatricalization in *Renaissance Self-Fashioning: From More to Shakespeare* (Chicago: University of Chicago Press, 1980), pp. 165–9; Goldberg discusses James I's trope of the king as actor, pp. 113–14. For an excellent re-assessment of nationalism and monarchical theatricality, see Paul Stevens, "Milton's Janus-faced Nationalism: Soliloquy, Subject, and the Modern Nation State," *JEGP*, 100.2 (2001): 264–8.

set on a stage, whose smallest actions and gestures, all the people gazingly doe behold" (43). Little attention, however, has been paid to the king's emphasis on nurture as a male political strategy. The king urges Henry to supply his own "nourish-milke" to his subjects so as "to win all mens hearts to a loving and willing obedience" (20). James thus makes the new primacy of nurture serve the old sovereignty: he can win hearts and minds as well as, more traditionally, force submission and compel obedience.

James's task thus is to project a seamless image of his sovereignty on the public stage: to integrate his public, paternal authority with his private, maternal love and nurture. But, while projecting integrity and absoluteness in gendered terms, James's double role as maternal–paternal sovereign also bears the potential to open up fault lines in the royal image, which the king must struggle to close up or cover over. James's prefatory letter to the English reader added to the 1603 edition provides a vivid demonstration of the king's struggles to strengthen the fragile connection between his strong, public paternal persona and his image of his loving and nurturing private maternal self. At the same time, however, his divine and rogynous sublimity threatens to lapse into monstrous hermaphroditic doubleness and division, thus shattering the royal image and scattering its broken shards. In his preface—and so even before the curtain goes up on the dramatic spectacle of royal sovereignty, for which his book serves as a theater—James insists on his absoluteness. He maintains that his public performance reveals his authentic inner self:

> Kings being publicke persons, by reason of their office and authority, are as it were set (as it was said of old) upon a publicke stage, in the sight of all the people; where all the beholders eyes are attentively bent to looke and pry in the least circumstances of their secretest drifts (5).

The king's outer "publicke" strong patriarchal rule and his inner nurturing maternal self are one and the same: indivisible and absolute.

James establishes this point through another favorite metaphor: the mirror. Through his emphasis on seeing and being seen and on his center-stage position, "in the sight of all the people; where all the beholders eyes are attentively bent to looke" (43), James offers himself as a mirror of majesty. Elsewhere in the text, his metaphor of the mirror allows him to project an image of never-ending Stuart dynasty, but in this instance the mirror serves a different function: it unites the king's highly visible public paternal presence with his hidden maternal self. Rather than replicate *sine finum* the royal image, this mirror discloses to the eyes of "all...beholders" the royal substance beneath: the king's nurturing maternal spirit. Like the mirror that Hamlet famously associates with the revelatory powers of acting, James's royal performance is "Not such a Mirror whereein you may see your own faces, or shadowes, but such a Mirror, or Christall, as through the transparantnesse thereof, you may see the heart of your King" (306). The royal actor possesses a "transparantnesse" (306) that allows the spectator to match paternal sovereignty with maternal nurture. Through his "transparantnesse," the king's public performance reveals that which otherwise cannot be shown ("that within which passeth show" [*Hamlet*, 1.2.85]): "the heart of your King" (306) or the king as "nourish-father."

This transparency, however, is double-edged. If it unites his inner maternal and outer paternal selves, it also renders James vulnerable in traditional female terms to the invasive gaze of undiscerning spectators "attentively bent to look and pry" (5). The king commands the attention of his whole audience, but the power of his imposing image nevertheless depends upon his audience's collective willingness to see him as he would like to be seen. James is well aware of the divisions and divisiveness of his audience: that "Hydra *of diuersly-enclined spectatours*" (9). He also knows that even the most skilled royal actor can be misjudged by a visually and conceptually impaired public: "Although a King be never so praecise in the discharging of his Office, the people, who seeth but the outward part, will ever iudge of the substance" (43). While "all the people gazingly doe behold" even the "smallest actions and gestures," they also may fail to connect "the outward part" to the "Kings inward intention" (43). The very same transparent performance that unites outward to inward, father and mother, also can divide the two states and tear apart the king's seamless image. The king's virtuoso balancing act—his integration of his public paternal and private maternal selves—thus does not always remain perfectly within his control. Although James attempts to incorporate all binaries within his all-encompassing, royal persona, he cannot always contain the potential for disintegration and self-division. Even as the king commands the attention of "all the people" (43) and so unites the public *en masse* through their collective focus on his royal person, James's absolute power nevertheless depends upon his subjects' discerning eyes and judgments, without which the royal center cannot hold: "praeiudged conceits will...breed contempt, the mother of rebellion and disorder" (43).

Once again, the rhetoric of maternal nursing does crucial political work for the king. He overrides his dependence as an actor on the people's applause and their questionable powers of discernment by cultivating a larger social need for his "nourish-milke" (24), a need that he alone can fulfill.[19] Equally important, however, through his proffers of "nourish-milk," the king ensures that his readers/audience will gain the interpretive

<hr />

[19] See Timothy Reiss, "Montaigne and the Subject of Polity," in *Literary Theory/Renaissance Texts*, ed. Patricia Parker and David Quint (Baltimore: The Johns Hopkins University Press, 1986), pp. 139, 137, 139–40.

powers they need to comprehend his character as he wishes it to be seen and understood. As already noted, Charity is typically allegorized as a nursing mother with a babe sucking milk from her breast. Time and again, James refers to his "Charitable reader" (4) or "Christian Reader" (7), who embraces love as the way to truth. Charity inspires insights that cross the divide between heaven and earth: it illuminates the ways in which ordinary, worldly objects, places, events, and characters have depths, histories, and purposes that exceed immediate apprehension or commonplace understanding. James's charitable readers (his metaphorically breast-fed children) thus can see not only his temporal power, but also his inner, invisible, abstract virtues. They recognize the "transparantnesse" (306) of the king's role on the public stage as a mirror of truth. Citing Luke 12, James writes:

> *that there* is nothing so couered, that shal not be reuealed, neither so hidde, that shall not be knowen; and whatsoeuer they have spoken in darknesse, should be heard in the light; and that which they had spoken in the eare in secret place, should be publickely preached on the tops of the houses (4).

In styling himself as a "nourish-father," or a male allegory of the nursing Charity, James not only attaches his readers to his nurturing breast, but he also fills them with the benevolence and beatitude they need to experience revelation, or, in this case, to connect what is "hidde" (the king's heart and maternal nurture) to what is "reuealed" (his male public persona and power). Even more importantly, however, the allusion to Luke underscores the coherence between the king's politically motivated evocations of the Apocalypse and his nurturing self-image. As a nursing father, James engenders charity in his readers—the very same charity they will need to envision him as he wishes himself to be seen: whole, perfect, and absolute, in the image of Christ as king of the salvific, post-apocalyptic millennial kingdom.

By appropriating Reformed connections among nurture, nature, and nation, James thus gains a considerable amount of symbolic political capital. As a nursing father, he not only adds new value and prestige to his dynastic authority and power, but he also dispels the shadow cast by his dead mother. By speaking the new language of nurture, the king intensifies the cult of monarchy: he centers all social relations on himself as sole provider of mother-love and nurture.[20] He also expands his rule into men's

---

[20] Protestant repression of Marian devotion further reinforced the king's role as cultic mother. See Frances Dolan, "Marian Devotion and Maternal Authority in Seventeenth-Century England," in *Maternal Measures*, 2000, pp. 282–91.

hearts; he pushes past unreliable, ritual displays of obedience to gain control over the subject's sealed-off interiority. By extending his nurturing breast to the people, the king fulfills his ambition to "to win all men's hearts to a loving and willing obedience" (20), affectively binding the people to himself through his "nourish-milk" (24). And, paradoxically, while mother-love intensifies the mysterious spectacle of James's monarchy, it also conveys the king's Protestant disdain for glittering surfaces. The nursing father is just one of the parts James plays in the drama of kingship, but this affecting role allows him, as a "mother," to nurture the heart and win the binding love and willing obedience of his son and subjects.

In *Basilikon Doron*, love and obedience, depth and surface, reformation and tradition, female and male, all fuse at the breast of the all-encompassing, nurturing king. As a nursing father, James aligns himself and his book with both the traditional *and* the reformed family. He embraces the new family's moral and spiritual function to nurture souls and to create affective bonds between parents and children. But, he also points the new family/nation back toward the old by placing the emphasis on shaping the inner character of children in service to the customary, genealogical *raison d'être* of the dynastic family. This fusion of old and new greatly expands the range of James's politic rhetoric: he can proclaim the freedom from the law that conscience and inward authority uniquely accord him; expand his royal governance into the hearts and minds of his son and subjects; and erase his own problematic maternal legacy by converting the tainted blood of his mother, Mary, Queen of Scots, into his own purifying maternal milk. In James's text, the trope of the nursing father underscores not only the marked differences but also the surprising overlaps within early seventeenth-century English Protestant culture between traditional, patriarchal displays of dynastic power and an anti-customary, maternal emphasis on nurture as crucial to constructing the new national subject.

EIKON BASILIKE: THE ROYAL IMAGE, CULTURAL MEMORY, AND THE DISCOURSE OF NURTURE

As in *Basilikon Doron*, in *Eikon Basilike*, the trope of the nursing father works to blur tradition and innovation, outer shows and inner truths. Neither a formal treatise nor a history of the English Civil War, *Eikon Basilike* provides a compelling narrative of the tumultuous events from the meeting of the Long Parliament in 1641 to the king's imprisonment in 1647 from the vantage point of Charles (and his publicists). It is "Charles I's most enduring and most powerful legacy to his own century and to history," as Kevin Sharpe observes.[21] The book also creates a verbal icon of the king that differs considerably from visual depictions of his royal person. As Elizabeth Skerpan Wheeler points out,

> In contrast to the earlier, carefully manipulated images of Elizabeth I, James I, and Charles I himself... *Eikon Basilike* appeared without an enveloping royal context... In the absence of direct royal control or effective government censorship, this image immediately developed an autonomous life, appropriated by readers and the book trade to create a publishing phenomenon.[22]

Royalist praise for the text greatly enhanced the book's marketing potential. *Eikon Basilike* hit the streets on the day of the king's execution. Within a month and half, approximately twenty English editions were printed, and by the end of 1649, that number reached thirty-five. Three Latin and seven Dutch editions appeared in the same year, extending the influence of the book well beyond England.

The unprecedented success and popular appeal of the book inspired counter-attacks from the opposition, most notably, from Milton in *Eikonoklastes*. Among other things, Milton and other anti-Stuart writers challenged Charles's authorship. Milton suspects that a "secret Coadjutor" (*CPW*, 3: 346) at least partly wrote the book; and, in his comments on the "Pamela prayer," the poet also argues that the king's seemingly sincere expressions of piety and his love for God and the people were in fact not only plagiarized but taken from a wholly inappropriate source: the "Heathen fiction" of Philip Sidney's *Arcadia* (*CPW*, 3: 362). The king's critics were quite right to question Charles's authorial authenticity, since hands other than his certainly played an important role in shaping *Eikon Basilike*. The text in fact was written by the Presbyterian divine, John Gauden, although a strong case can be made for dual authorship: that Gauden based his text on meditations which the king began to write during his confinement at Carisbrooke Castle in 1647. As Philip Knachel points out, "twenty-six of the twenty-eight chapters of *Eikon Basilike* relate to events either connected with his stay there or prior to it."[23]

Still, even if we see Charles as co-authoring or, at the very least, as supervising Gauden's writing of the book, *Eikon Basilike* cannot be read

---

[21] Kevin Sharpe, "The Royal Image: An Afterward," in *The Royal Image: Representations of Charles I*, ed. Thomas N. Corns (Cambridge: Cambridge University Press, 1999), p. 288.

[22] Elizabeth Skerpan Wheeler, "*Eikon Basilike* and the Rhetoric of Self-Representation," in *The Royal Image: Representations of Charles I*, p. 122.

[23] Philip A. Knachel, "Introduction," *Eikon Basilike: The Portraiture of His Majesty in His Solitudes and Sufferings*, ed. Philip A. Knachel (Ithaca: Cornell University Press, 1966), p. xxx. All quotations from *Eikon Basilike* are taken from this edition and noted in the text.

simply as Charles's spiritual autobiography; nor does it provide a conventional icon of the king. As Skerpan Wheeler notes, "there is no single, unified, 'official' version of the text."[24] The king's image is a shifting and malleable composite comprised from the book's many English editions, including a verse edition set to music, its Latin, Dutch, and French translations, and its addendae and accretions. On March 15, 1649, the printer, William Dugard, published an edition of *Eikon Basilike* with new supplementary materials: four prayers attributed to the king; a letter written from Charles, Prince of Wales, to his father; narrative accounts of the king's last moments with and parting words to his children; an epitaph on Charles's death; and a collection of apothegms derived from editions published earlier in the year. These additions proved so popular that they were immediately inserted into unsold copies of the text and, subsequently, they were incorporated routinely into almost all later editions.

In this section, I wish to demonstrate how the new rhetoric of maternal nursing strengthens Charles's construction of his royal image in *Eikon Basilike*. Much attention justly has been focused on how, in his book, the king casts himself as a pious martyr, willing to die for a noble cause. But, scant attention has yet been paid to the ways that Charles's emphasis on his self-sacrificing love for his people gains power from the period's revaluations of maternal nurture. As we shall see, nurture is a defining aspect of Charles's official self-image even before he ascends to the throne, and then again during his long Personal Rule in the 1630s. Linking maternal nurture to paternal strength helps Charles to offset the martial, Arthurian image that, as Thomas Corns argues, is associated with Prince Henry, who died in 1612 before ascending to the throne. During the years of Charles's Personal Rule, the king's composite image as nurturing and strong, maternal and paternal reinforces his singularity and absoluteness in gendered terms. In Van Dyck's famous *Equestrian Portrait of Charles I* (1637; National Gallery, London), the king is dressed in armor and is wearing the Medallion of a Garter Sovereign (see figure 11).

His face bears a serene expression, which, at first glance, seems at odds with his military posture and chivalric attire. Charles holds a commander's baton, which is mostly hidden; at the same time, however, his spirit bends (in a "female" posture) toward a higher authority. Commanding and yielding, masculine and feminine, Van Dyck's Charles I epitomizes the sublimity of kingship. I argue that Charles and his apologists attempt to recreate a similar kind of sublime androgyny in *Eikon Basilike* by portraying the king as a nursing father. This is not to say that Charles does

[24] Elizabeth Skerpan Wheeler, "*Eikon Basilike* and the Rhetoric of Self-Representation," p. 123.

Fig. 11. Anthony van Dyke, *Equestrian Portrait of Charles I*, 1637.

not depict himself as a Christ-like martyr in his book, but just to identify a few more hitherto unrecognized layers in the king's pious depiction of his adversities. With its changeable, unstable qualities, *Eikon Basilike* is ideally suited to providing a multifaceted, shifting portrait of the king. In his book, Charles appears in more than one guise at the same time without undermining any aspect of his malleable old–new, past–present, male–female image. The king is a Christian martyr *and*, like his own father, he too is a nursing father (a mighty patriarch and a pious-and-tender "mother") all at the same time.

   In part, Charles's proffers of nurture and love to his subjects seem designed to correct regicidal depictions of the king as a hard-hearted tyrant and monstrous father, incapable of feeling compassion for his own people. John Cook, in *King Charls his Case* (1649), links Charles's slippery politics and acts of treachery in the bloody Irish Rebellion of 1641 to the king's lack of paternal feeling: he would "raise War on his own children."[25] Cook maintains that Charles selfishly pursues his own personal pleasures at the expense of the people; he turns the royal hunt (both James's and Charles's favorite sport) into a metaphor of the king's self-indulgence and his savage disregard for his own subjects. Cook writes that the king made "the People...his Venison to be hunted at his pleasure."[26] Charles's lack of compassion and his craven appetites prove that he is "that great *Nimrod*" of his age and not a Moses, the nursing father of his people.[27] In *The Tenure of Kings and Magistrates* (1649), Milton deploys a similar argument to justify the regicide: the king became the "Tyrant *Nero*" when, with a group of armed men, he made a charge against the House of Commons in an attempt to arrest five Parliamentary leaders of the opposition on January 4, 1642 (*CPW*, 3: 439–40). Like Cook, Lucy Hutchinson depicts the king as motivated entirely by self-interest and as completely devoid of feeling for the needs of his subjects. For Hutchinson, as for both Cook and Milton, Charles is a tyrant, not a king: he was "the most obstinate person in selfewill that ever was, and so bent upon being an absolute uncontrowlable Soveraigne that he was resolv'd either to be such a King or none."[28] The king is absolute only in his ungovernable selfishness. His cold-hearted drive to achieve absolute sovereignty at the expense of his subjects' well-being proves that he is "none," since "to be such a King" is not to be a king at all, but rather an "uncontrowlable" tyrant. Charles's kingship and kingdom are null and void—politically "dead."

Nevertheless, one of the king's harshest critics, Lucy Hutchinson, looking back several years later, would praise Charles for reforming the Jacobean court. Under Charles, "The face of the court was much changed in the change of the King, for King Charles was temperate, chaste and serious, so that the fools and bawds, mimics and catamites, of the former court [of the Comus-like James] grew out of fashion."[29] Hutchinson's characterization of Charles as an agent of moral reform finds an unlikely analogue in Jonson's *Pleasure Reconciled to Vertue*, the 1618 masque celebrating Charles as the new Prince of Wales. Jonson's masque is intriguing because it closely relates Charles's potential power to civilize his father's

[25] John Cook, *King Charls his Case* (London, 1649), p. 39, cited in David Loewenstein, "The King Among the Radicals: Godly Republicans, Levellers, Diggers and Fifth Monarchists," in *The Royal Image: Representations of Charles I*, ed. Thomas N. Corns (Cambridge, Cambridge University Press, 1999), p. 103.

[26] James describes "the hunting, namely with running hounds" as "the most honourable and noblest sorte [of sport] thereof" (*BD*, 48).

[27] Cited in Loewenstein, p. 103.

[28] Lucy Hutchinson, *Memoirs of the Life of Colonel Hutchinson*, ed. James Sutherland (London: Oxford University Press, 1973), p. 47.

[29] Lucy Hutchinson, *Memoirs of the Life of Colonel Hutchinson*, ed. James Sutherland (London, Oxford University Press, 1973), p. 46.

court to his maternal capacity for nurture. Nurture not power breeds virtue. Whereas Hutchinson provides us with backward glance at the striking contrast between the "sober, chaste, and serious" Charles and the licentious James, Jonson looks forward to Charles's moral reformation of the Jacobean court. Unlike Hutchinson, however, Jonson closely associates the king's capacity for nurture with his moral agency and his role as national redeemer.

In *Pleasure Reconciled to Vertue*, masquers led by Charles, the new Prince of Wales, emerge from a cleft in "the Mountaine ATLAS."[30] The decadent regime of "the god Comus here, the *Belly-god*" (l: 481) has just ended with the disappearing antimasque "of *Pigmees*." Mercurye invites the audience to look forward to a future era, in which "One...of the bright race of *Hesperus*" (11: 204–5) shall reign. With Charles's ascent to the throne, misrule will yield to order, appetite to reason, and, as the title proclaims, pleasure will be reconciled to virtue. As Corns points out, *Pleasure Reconciled to Vertue* (completed in 1618) formally "resembles quite closely *Oberon, the Fairy Prince*, Prince Henry's Christmas masque of 1611." But whereas in Henry's masque, "a chivalric—indeed Arthurian—aesthetic obtains," in Charles's masque, "the Arthurianism of *Oberon* gives way to a decidedly Horatian aesthetic, which relegates martial nature to civilizing nurture." Corns's main objective is to demonstrate that Charles, as the newly created Prince of Wales, distinguishes himself from his late, elder brother by containing and suppressing the Arthurianism that Prince Henry might have revived. Pointing Corns's observation in a somewhat different direction, I would add that the Horatian impulse in *Pleasure Reconciled to Vertue*, which "relegates martial nature to civilizing nurture," also allows Jonson to anticipate the cultural agenda for Charles's future court. Maternal nurture as well as a martial nature will be a governing principle of the Caroline era.[31]

This precise point is made in the Song that accompanies the moment in Jonson's masque when Mount Atlas opens up to reveal the masquers inside, led by Charles. Mighty, "*aged*," and decidedly masculine, the mountain, in the very first two lines, is compared to "an old Man, his head &, beard all hoary & frost" (ll: 1–2). In the Song, however, the mountain's old, wintry, and paternal qualities are superseded by his maternal fruitfulness and nurture. If he is "an old Man," he is also a childbearing and nurturing mother. The mountain is commanded to "*open*

---

[30] Ben Jonson, *Pleasure Reconciled to Virtue*, in *Ben Jonson*, ed. C.H. Herford and Percy and Evelyn Simpson, 7 (Oxford: Clarendon, 1941), p. 479. Line references are taken from this volume and noted in the text.
[31] Thomas N. Corns, "Duke, Prince and King," in *The Royal Image: Representations of Charles I*, ed. Thomas N. Corns (Cambridge: Cambridge University Press, 1999), pp. 4, 5.

then thy lap[and from thy beamy bosom, strike a light," so that those inside, who, as stated in the stage directions, come "forth from the Lap of the Mountaine" and be nurtured:

> *may read in thy misterious map:*
> *all lines*
> *and signes*
> *of roial education and the righ*[t]
> (ll: 221–3)

With Prince Charles, the "glory of the West" (l: 192), leading the way, the masquers emerge from the womb-like interior of the mountain eager for intellectual sustenance. In this allegorical moment of rebirth, "*roial education and [birth] righ[t]*," "*skill*" and "*will*," fuse together. The Song describes those who pass through the (labial) cleft in the rock as "*borne to know*," to lap up social knowledge from the breast of the nursing father. In the new era, "*they who are bred/within the hill/of skill/may safely tread/ what path they will*" (ll: 230–4). Just as pleasure will be reconciled to virtue, so too will nature (birthright) be reconciled to maternal nurture (moral education) in the future Caroline era.[32]

Whereas Jonson emphasizes Charles's role as a nurturing ruler, Renold Elstrack's portraits of the king (the first showing Charles in his teens, about ten years before his accession, the second as king in 1625) as well as the king's coinage highlight Charles's Arthurian character. John Peacock points out that many images of Charles from the 1620s and 1630s, featured variously on coins, medals, statues, engravings, and paintings, emphasize his monarchial strength and power.[33] As Peacock argues, these images tend either to depict the king on horseback, thus emphasizing his chivalric persona, or they present a realistic portrait (rather than a generalized royal image as was the standard medieval practice) with imperial associations. As I have already noted, Van Dyck, in his famous portrait *Equestrian Portrait of Charles I*, idealizes *both* aspects of the king. This painting helped to make the king one of the most recognizable of all English monarchs.[34] In Van Dyck's paintings of the 1630s, Charles's motherly love and paternal strength complement and constitute one another. Van Dyck's Charles is "the monarch of the Personal Rule, and Van Dyck's

---

[32] Lucy Hutchinson, *Memoirs of the Life of Colonel Hutchinson*, ed. James Sutherland (London, 1973), p. 46.

[33] My discussion of Charles's multi-media circulation of his royal image is indebted to John Peacock, "The visual image of Charles I," in *The Royal Image: Representations of Charles I*, ed. Thomas N. Corns (Cambridge: Cambridge University Press, 1999), pp. 176–239.

[34] See Kevin Sharpe, "The Royal Image: An Afterward," in *The Royal Image: Representations of Charles I*, ed. Thomas N. Corns (Cambridge: Cambridge University Press, 1999), p. 288.

function is to picture his power as sympathetically as possible."[35] Van Dyck, the most important of the many continental artists, including Rubens, attracted to the Caroline court, represents the king as the king ideally saw himself and as he would have others see him—as both a chivalric hero *and* as a nurturing father to his people.[36]

With this brief history of Charles's verbal and visual portraiture in mind, let us turn now to the famous frontispiece by William Marshall to *Eikon Basilike*. The portrait is designed to evoke maximum sympathy from the book's readers. Rather than a chivalric hero on horseback, the icon represents the king as the exemplar of passive suffering. Charles is shown in a basilica, kneeling in prayer. His earthy crown is at his feet, but his eyes are fixed on a heavenly crown, which radiates light, life, and the after-life. In his hand, the king holds a crown of thorns, symbolic of his Christ-like agony and his redemptive acts of self-sacrifice. At the left, a massive rock stands firm despite the tempestuous waters that surround it; a palm tree bears heavy weights. The crown of thorns and the rock both emblematize the king's patient bearing of his burdens and his triumphant steadfastness in the face of injustice, persecution, and anarchy. As already noted, much emphasis has been placed on the effectiveness with which Charles performs the role of Foxean martyr in his book.[37] As David Loewenstein observes, Charles "appears as a martyr, who has suffered terrible afflictions at the hands of violent and malicious enemies."[38] As we shall see, however, Charles solidifies his image as a Christ-like martyr elsewhere in the text through the reformed discourse of maternal nurture. Time and again, he highlights the noble sacrifices he makes to nourish and care for his subjects. He remains their Christ-like nursing father, despite his own terrible afflictions. Charles also exploits the affective power he cultivates as a nursing father to revive his readers' memories of his signature, compound image as the nurturing *and* chivalric sovereign of the Personal

---

[35] Peacock, "The visual image of Charles I," in *The Royal Image: Representations of Charles I*, p. 228.

[36] For a foundational and still very rich and current interpretation of this famous painting, see Roy Strong, *Van Dyck: Charles on Horseback* (London: Allen Lane, 1972).

[37] For commentary on Charles I and martyrdom, see John Knott, "'Suffering for Truth's Sake': Milton and Martyrdom," in *Politics, Poetics, and Hermeneutics in Milton's Prose*, ed. David Loewenstein and James Turner (Cambridge: Cambridge University Press, 1990): 153–70; Laura Lunger Knoppers, *Historicizing Milton: Spectacle, Power, and Poetry in Restoration England* (Athens: University of Georgia Press, 1994), Chapter 1; and Robert Wilcher, *The Writing of Royalism, 1628–1660* (Cambridge: Cambridge University Press, 2001), pp. 276–86. None of these studies addresses either the gendered implications of the king's self-image as a heroic/Christ-like martyr or the ways that Charles intensifies the affective power of his Christ-like sufferings by deploying the new rhetoric of nurture and the medieval image of Jesus as a maternal nurse.

[38] David Loewenstein, *Milton and the Drama of History: Historical Vision, Iconoclasm, and the Literary Imagination* (Cambridge: Cambridge University Press, 1990), p. 53.

Rule—an image belied by the king's actual imprisonment in Carisbrooke Castle in 1647, where he is supposed to have composed his book. Charles's complex efforts to spin his royal image to maximum political advantage requires, impossibly (or so it would seem), that, on the one hand, he deflect attention away from his actual adversities in order to conjure up the memory of his sublime image as the maternal–paternal monarch of the 1630s, and, on the other, that he focus exclusively on his current trials and tribulations in order to depict himself as a heroic martyr. The discourse of nurture allows the king to meet both of these mutually exclusive objectives at the same time.

From the very start, *Eikon Basilike* highlights Charles's self-sacrificing devotion and the plenitudinous love he has for his people: he "was indeed sorry to hear with what partiality and popular heat elections were carried in many places;" he nevertheless, "knowing best the largeness of my own heart toward my people's good and just contentment, pleased myself most in that good and firm understanding, which would hence grow between me and my people" (*EB*, 4). Rather than increase partiality by encouraging factionalism, the king will unite his nation through love and nurture by extending "the largeness of my own heart toward my people's good and just contentment" (*EB*, 4). Although outward circumstances have dramatically reduced his status, stature, and traditional masculinity, the "largeness of my own heart" allows Charles to replenish his depleted male majesty in affective, inward, private, and maternal-coded terms. Through love and nurture, he will restore his lost monarchical grandeur in new abstract and internal sites: the hearts, minds, and memories of his subjects. Charles is a self-sacrificing martyr, who remains wholly devoted to "my own children's interests"; he seeks only to "preserve the love and welfare of my subjects," despite his own personal sufferings (*EB*, 4). As a nurturing father, he also gains new symbolic power by fostering his subjects' love, which will "hence grow" in spirit and in memory. As suggested by his spiritual rendering of the organic metaphor of growth, Charles appeals to the new abstract terms of identity and social worth to compensate for the devaluation of the Stuart dynasty's traditional, organic measures of power, social position, and authority, as well as for his own political and material losses. Outer loss converts to inner gain. "I cared not," writes Charles, "to lessen myself in some things of my wonted prerogative, since I knew I could be no loser if I might gain but are compense in my subjects' affections" (*EB*, 4).

This equation between inner gain and outer loss, nurture and power, the new mother and the (memory of) the old father is illustrated in the three narratives included in Dugard's addenda, which feature tender scenes of the king *en famille*. Here as elsewhere in his book, Charles is depicted as selflessly devoted to his children and, by implication, to his subjects as well. Attributed to his daughter, Elizabeth, these narratives showcase Charles in the reformed role of tender, nurturing father. At the same time, however, they also attempt to establish that, despite his impending execution, the king's traditional role as patriarchal progenitor of an unbroken dynastic line remains intact:

> His children being come to meet him, he first gave his blessing to the Lady Elizabeth and bade her remember to tell her brother James whenever she should see him that it was his father's last desire that he should look no more upon Charles as his eldest brother only but be obedient unto him as his sovereign, and that they should love one another and forgive their father's enemies. Then, said the King to her, "Sweetheart, you'll forget this." "No," said she, "I shall never forget it while I live," and pouring forth abundance of tears promised him to write down the particulars. (*EB*, 192)

Affecting and sentimental, this narrative is also highly motivated, as the king's critics, especially Milton, make clear. In *Eikonoklates*, Milton accuses the king of exploiting his readers by courting their affections with a pretense of piety and false tears. Charles's expressions of love, especially the exquisite tenderness he displays when he calls his daughter "Sweetheart," are poignant, but they also do a great deal of political work. Through his affecting performance as a tender and caring father, Charles strives to correct his enemies' depictions of his unnatural lack of feeling. He asks his children not only to love one another but also to forgive his enemies. Rising above the fray and exhibiting his transcendent impartiality, Charles assumes the moral high ground, where he reclaims in spirit his natural rights as dynastic king. He makes a moving appeal to his younger son, James, to honor his elder brother's dynastic legacy—an appeal made all the more poignant and emotionally powerful by the fact that "it was his father's last desire" (*EB*, 192). Charles is a tender, Christian father, who nurtures all, friends and enemies alike, and he is a dynastic progenitor with a healthy line of male heirs. He is an old–new, male–female, procreative–nurturing patriarch. He thus figuratively remains whole and absolute, despite the actual dissolution of his kingship and dismantling of his royal sovereignty.

Charles's role as nurturing father thus is the key to the Dugard narrative's emphasis on ensuring that the Stuart dynastic line remains in tact, as his poignant appeal to his younger son demonstrates. This same role is equally crucial to the story's preoccupations with crafting, preserving, and protecting the king's image, as dramatized by the sentimental dialogue between Charles and his daughter, Elizabeth. Elizabeth responds to her father's fears that she will forget him with a decisive "No." "I shall never forget it while I live," she proclaims. In tears, she pledges to "write down the particulars" (*EB*, 192). Preservation of and control over the king's memory depend upon "the particulars" that Elizabeth chooses to write down as an expression of her love for her father—the same "particulars" that the book presents to its readers. Just as Charles is affirmed as his father's heir, so Princess Elizabeth is granted custody of her father's memory. If Charles II dynastically replicates Charles I, Elizabeth also replicates her father, but by a different means: by ensuring that his life story is recorded for posterity. Her affecting remembrance will move readers to act on their own love for their king and father, such that they too will become his metaphoric "daughters" and loving custodians of his memory. Like Elizabeth, they will "write down the particulars" (*EB*, 192) and remember Charles as he wishes to be remembered. Both the old natural measures of dynasty and the new linguistic and abstract measures of English nationhood lend value and worth to the king's carefully crafted royal image, to be preserved forever in the hearts of his children/subjects.

Whereas the private realm of the family models the ways that the king's idealized memory will be preserved in Dugard's narratives, the public realm inspires a similar kind of memory work in Chapter 5, "*Upon His Majesty's Passing the Bill for the Triennial Parliaments.*" Here, Charles repudiates Parliament's accusation that he soon regretted the assent he gave on February 16, 1641 to the Triennial Act, which mandated a calling of Parliament if Charles did not do so three years after dissolving the previous Parliament: "I could not easily nor suddenly suspect such ingratitude in men of honors that, the more I granted them, the less I should have and enjoy with them." As in the supplementary narratives, the king's conversion of loss into gain structures his depiction of his nurturing relationship to his people. Charles notes that, while unrecompensed by Parliament for the generous concession he made in passing the Triennial Act, ultimately he will gain rather than lose if he wins his subjects' hearts: if they return in kind the love he bears for them. "Nor do I doubt," Charles writes, "but that in God's due time the loyal and cleared affections of my people will strive to return such retributions of honor and love to me or my posterity as may fully compensate both the acts of my confidence and my sufferings for them" (*EB*, 21). Deploying economic metaphors of compensation and repayment, Charles depicts his love and concern for the well-being of his people as an investment that will pay dividends "in God's due time," in the form of the "honor and love" through which the people will remunerate the king for selfless devotion to them as a kind of Jesus as nurse. In this way, he paints his own posthumous image.

Through the love reciprocally exchanged between Charles and his subjects, to their mutual benefit, the king shields his royal image and dynastic legacy from the assaults made by his detractors. Just as his subjects open their hearts to their king, so the king opens his heart to his subjects. In exchange for preserving Charles's idealized image and protecting his memory, readers are offered free-and-open access to his most intimate, innermost self. The *"truth and uprightness of my heart"* is the focus of the king's prayer at the end of Chapter 6 (*EB*, 23). Tying his dedication to conscience to the disclosure of his heart, Charles links his deepest moral convictions to his most intimate feelings, thus revealing his exemplary inner character (which, he charges, is suppressed by his enemies) to his readers. In Chapter 6, he writes that no man shall "gain my consent to that wherein my heart gives my tongue or hand the lies; nor will I be brought to affirm that which in my conscience I denied before God" (*EB*, 21). By publicly displaying his private commitment to the dictates of his conscience, Charles breaks up the foundational Puritan equation among conscience, dissent, and liberty, and he re-identifies conscience as a royal prerogative. He also directs his subjects to judge him from the inside out and not, as do his enemies, from the outside in. Rather than evaluate him according to his dubious political actions, Charles's subjects will recognize him as a conscience-driven, self-denying, and nurturing monarch: *"My people can witness how far I have been content for their good to deny myself in what Thou has subiected to my disposal"* (*EB*, 24). In commenting on Parliament's failure properly to appreciate his willingness to support the Triennial Bill, Charles evokes the higher authority of the deity, "Whose all-discerning justice sees through all the disguises of men's pretensions and deceitful darknesses of their hearts" (*EB*, 23). The king's readers, in accordance with highest justice, must likewise attempt to "see through the disguises of men's pretensions" or, in this instance, disregard Parliament's denunciation of the king's support for triennial parliaments as hypocritical and untrustworthy.

Charles thus needs to create a reading public that not only can see him as he wishes to be seen, but that also can help him to recall and re-present those aspects of his royal character obliterated by his actual circumstances after 1647. Together, both the king and his readers will perfect the royal image in ideal terms by adding the memory of the king's dynastic power and authority to the perfect love and devotion that Charles extends to his people through his book. That he addresses his readers almost as a specter from beyond the grave makes his appeal to (fabricated) memory all the more urgent and undeniable. Not unlike the Ghost of Hamlet's father, Charles wants his subjects to avenge his foul murder and to remember him as both the traditional monarch and the loving father he once was. The idealized king celebrated in *Eikon Basilike* is far more affecting and tenderly maternal than Charles in his actual, demystified, and disgraced circumstances. In addition to winning his readers' sympathies as a Christian martyr, Charles focuses on his Christ-like desire to nurture so that, out of political expedience, he can emotionally attach his readers to his perfect performance of paternal love and nourishing kingship—and erase the actual failure of his monarchy and the compromising of his public character.

Charles's deployment of the rhetoric of nurture as a way to preserve his legacy and sustain his power despite his impending execution finds additional expression in Chapter 3, "*Upon His Majesty's Going to the House of Common.*" Referring to a scandalous event that occurred on January 4, 1642, Charles asks his readers to understand that his plan to arrest John Pym, John Hampden, Arthur Hesilrige, Denzil Holles, and William Strode (who were five of the principal leaders of the Parliamentary opposition) was motivated by his desire for justice. It was not, as both "indifferent men" and "many of my friends" represented it, "a motion rising from passion than reason, and not guided with such discretion as the touchiness of those time required." (The five opposition leaders in fact were tipped off and made their escape before Charles arrived at the Commons.) Charles never names names and so never accords his enemies status as discrete or autonomous individuals; instead, he depicts them only as an unruly and highly prejudiced, but generalized faction. He charges that "these men knew not the just motives and pregnant grounds with which I thought myself so furnished" (11). Unlike "these men," who neither see nor understand Charles's inner motives, the king's beloved and loving subjects can see their monarch's loving intentions, which are invisible to his enemies, fully reflected in his outward actions—the same actions which, according to Charles's detractors, make the king's political opportunism and empty theatricality transparent. If, by deploying the rhetoric of mother-love and nurture, Charles can seduce the people into concentrating on his inner rather than his outer man, then he can ensure that his charging of the Commons will be remembered as a crusading expression of his Christ-like love and devotion to his people and nation rather than as an unprecedented act of dubious legality (which in fact only increased opposition to Charles in both Houses of Parliament). Instead of adhering to the misperceptions and inadequate understanding of his blinded enemies, the discerning charitable readers of the king's book will see and interpret their nursing father's public performance at the Commons as a principled and heroic stand, a stand motivated equally by his heroic valor, pious devotion to God, and his love and compassion for his subjects.

The discourse of mother-love and nurture thus gives Charles a considerable edge in the battle against his enemies for representational control over the meaning and memory of his life and death. Through his repeated proffers of nurture and self-sacrificing love, the king wins not only the hearts and minds of his people but also their undying attachment even after his death. By contrast, in his powerful point-by-point attack against Charles's book in *Eikonoklastes*, Milton berates the English people for allowing themselves to be hoodwinked by their unscrupulous and manipulative monarch: "The People, exorbitant and excessive in all thir motions, are prone afttimes not to a religious onely, but to a civil kinde of Idolatry in idolizing thir King" (*CPW*, 3: 343). They are a "credulous and hapless herd, begott'n to servility, and inchanted with these popular institutes of Tyranny" (3: 601); "what a miserable, credulous, deluded thing that creature is, which is call'd the Vulgar" (3: 426)—"an inconstant, irrational, and Image-doting rabble" (3: 601). Rather than feed the popular appetite as does Charles, Milton appeals to the reason of the virtuous few: those fit readers capable of seeing and knowing the king as he is in fact rather than in the empty spectacle of royal sovereignty presented in *Eikon Basilike*. Milton maintains that Truth cannot be disguised, but is sent out

> in the native confidence of her single self, to earn, how she can, her entertainment in the world, and to finde out her own readers; few perhaps but those few, such of value and substantial worth, as truth and wisdom, not respecting numbers and bigg names, have bin ever wont in all ages to be contented with (*CPW*, 3: 339–40).

To highlight the duplicity that motivates Charles's royal performance, Milton exposes the seams that fail to hold the king's inner and outer selves together. Charles's revelation of his private, pious self is mere pretense and false theatricality. The crisis of conscience that Charles professes to experience after giving his assent to the bill of attainder (passed by the House of Commons on April 21, 1641), which granted Parliament legal permission to execute the king's close associate and principle advisor, Thomas Wentworth (Earl of Strafford), is a "shew" and "a subtle dissimulation" (3: 372). The king's poignant expressions of love for his people and his heartfelt concern for their well-being also lack substance and authenticity. The king should thus be judged by his actions alone. "Stage-craft," writes Milton, "will not doe it" (3: 530).

Milton's disdain for the affective power and popular appeal of the love and nurture that the king extends to his subjects as their nursing father ultimately undermines his case against Charles's manipulative theatricality, despite the forcefulness of the poet's arguments. Elsewhere in his writings, however, Milton demonstrates that he too can turn the discourse of mother-love and nurture to optimal political advantage, as we shall soon see. But, first, we need to understand why Cromwell appropriates the Stuart image of the nursing father and how the discourse of maternal love and nurture helps him to pry England loose from its dynastic antecedents.

"LOVE ALL, TENDER ALL": NURTURE AND NATIONAL UNITY IN CROMWELL'S PUBLIC SPEECHES

As in *Basilikon Doron* and *Eikon Basilike*, the language of maternal nurture in Cromwell's speeches speaks to the new importance of affect and feeling as criteria for determining social identity and status and for cultivating a sense of national community. But, whereas James's self-representation as a "nourish-father" exhibits the *sui generis* unity of his royal person, and whereas Charles's rhetoric of maternal–paternal love and nurture helps him to secure his monarchical integrity, even moments before his head is severed from his body, Cromwell's nursing fatherhood is reconfigured as selfless, civic-minded, maternal tenderness and nurture and then deployed to idealize a politically expedient vision of balance between central and local authority, between state power and individual conscience.

That balance was rather precariously maintained during the civil wars, especially after Cromwell's and Fairfax's decisive victory over King Charles at Naseby when the united front forged between Parliament and the New Model Army began to come apart. From the end of April to the beginning of January 1647, Parliament came to be dominated by conservative MPs "who have so much malice against the Army as besots them," as Cromwell writes in a letter to General Thomas Fairfax, dated March 11 of that year.[39] Angered by the Parliamentary resolution to disband the Army without paying arrears of soldiers' wages, creating pensions for widows and orphans, or passing an act of indemnity for illegal actions committed by soldiers while under Parliamentary command, among other grievances, the Army attempted to break rank with its officers, many of whom

[39] Quoted in Christopher Hill, *God's Englishman: Oliver Cromwell and the English Revolution* (London: Penguin Books, 1990, first published 1972), p. 81. Other excellent biographies of Cromwell include: C.H. Firth, *Oliver Cromwell and the Rule of Puritans in England* (London: Putnam, 1901); Barry Coward, *Oliver Cromwell* (London and New York: Longman, 1991); Peter Gaunt, *Oliver Cromwell* (Oxford: Blackwell, 1996). See also John Morrill, ed., *Oliver Cromwell and the English Revolution* (London and New York: Longman, 1990). Excellent histories of mid-seventeenth-century England include G.E. Aylmer, *Rebellion or Revolution?: England 1640–60* (New York: Oxford University Press, 1986); Derek Hirst, *Authority and Conflict: England 1603–1658* (London: Edward Arnold, 1986); Christopher Hill, *Puritanism and Revolution: Studies in Interpretation of the English Revolution of the Seventeenth Century* (London: Secker, 1958) and *The World Turned Upside Down: Radical Ideas during the English Revolution* (New York: Viking, 1972).

came to sympathize with their soldiers' demands. Christopher Hill describes the Army's attempt at revolt as marking "a turning point in Oliver Cromwell's career."[40] Fearful that the Army might be crushed by Parliament, lose its strength as a unified military and political force through its own internal battles, or fall under the control of the more radical Leveller party, Cromwell tied his political future exclusively to the Army, dedicating himself to preserving its unity at all costs. Sir John Berkeley, the Royalist with whom Cromwell and Ireton attempted to negotiate openly and in secret with the king, and who in 1656 came under suspicion of Charles II for communicating with Cromwell, writes that: "After Cromwell quitted the Parliament, his chief dependence was on the Army, which he endeavoured by all means to keep in unity."[41]

Cromwell, in Berkeley's negative description, is something of a devilish Machiavel: for Berkeley, "all means," good or evil, equally serve Cromwell's morally irresponsible quest for unity. For his sympathizers, however, Cromwell's all-out pursuit of unity reflects his virtuous dedication to Machiavelli's science of power. This is the Cromwell who ambiguously triumphs as the new-modeled Caesar in Marvell's *An Horatian Ode Upon Cromwell's Return from Ireland.*[42] Marvell celebrates the Lord General as a heroic, Machiavellian man of virtú, who topples traditional structures and breaks apart received moral categories in his "restless" and "cease"-less drive to assert his power with complete indifference to both "The Emulous" (l: 18) on "his own Side" (1: 15) and the Stuart "Enemy" (l: 18).[43] The inexorable natural and historical forces that drive Cromwell's self-activated ambitions are too enormous to be confined to any one side. Instead, they serve the ineluctable totality, perfect unity, and ideal community to which Cromwell appeals time and again in his speeches at different stages of his career. As we shall soon see, Cromwell naturalizes this innovative, category-bursting unity through the language of maternal love and nurture.

[40] Christopher Hill, *God's Englishman*, p. 83.

[41] Hill, p. 86.

[42] Norbrook observes that in both the *Horatian Ode* and *The First Anniversary of the Government under His Highness the Lord Protector*, Marvell represents Cromwell as "a Machiavellian legislator, one who is able to return the state to its best principles by decisive innovation." See *Writing the English Republic*, p. 342. Joseph Anthony Mazzeo was the first to note the Machiavellian character of Marvell's Cromwell in "Cromwell as Machiavellian Prince in Marvell's "Horatian Ode," in *Renaissance and Seventeenth-Century Studies* (New York: Columbia University Press, 1964), pp. 183–208. Blair Worden discusses Marvell's debt to Machiavelli in "Andrew Marvell, Oliver Cromwell, and the *Horatian Ode*," in *Politics of Discourse*, ed. Kevin Sharpe and Steven N. Zwicker (Berkeley: University of California Press, 1987) pp. 162–8.

[43] Andrew Marvell, *The Poems and Letters of Andrew Marvell*, ed. H.M. Margolouth, 2 vols. (Oxford: Clarendon, 1971).

In Marvell's Cromwell poems, Cromwell's quest for unity by all means reflects a Machiavellian vision of power politics and innovation.[44] As J.C. Davis observes, Cromwell's moral relativism also can be traced to his profound skepticism about both philosophical and political systems and formal church doctrine, ritual, and liturgy. For Davis, Cromwell's desire to impose unity by all means and his indifference to law, reason, and historical precedent stems from his radical uncertainty about human agency, artifacts, and institutions.[45] Such skepticism was clearly at work at the Putney debates in 1647, when Cromwell condemned the Leveller constitution as empty rhetoric. "I am," he maintains, "persuaded in my heart" that it is not political platforms, proposals, or petitions that "tends to uniting us in one." "It is not enough," he asserts, "for us to propose good things."[46] The same radically disenchanted view of externals also inspires his governing moral conviction that reformation could proceed only by adhering to the "free way" rather than the "formal way" to war and social change.[47] The free way to political unity had been Cromwell's way from the start. In 1645 Cromwell wrote to Speaker Lenthall concerning the Army:

> Presbyterians, Independents, all had here the same spirit of faith and prayer.... They agree here, know no names of difference; pity it is it should be otherwise anywhere. All that believe have the real unity, which is most glorious because inward and spiritual....[48]

Not unlike Drayton's goddess-mother of Great Britain, Cromwell is the consolidating "nursing mother" of the Army, in which "real unity" for all is perfectly realized. By contrast, in all other places, "otherwise anywhere," the "names of difference," thwart the sameness of spirit that emanates from within, "which is most glorious because inward."

In striking but hitherto unrecognized ways, the period's new preoccupations with motherhood, childhood, the family, and other domestic matters also inspire the Lord General's anti-formalism. Marvell's *An Horatian Ode* represents Cromwell's free way through received traditions

---

[44] Norbrook, p. 342.

[45] J.C. Davis, "Cromwell's Religion," in *Oliver Cromwell and the English Revolution*, ed. John Morrill (London and New York: Longman, 1990), p. 1: Longmans, Green, and Co., 1891–1901), I. 440-481-208.

[46] *The Clarke Papers*, ed. C.H. Firth (London: Putnam, 1901).

[47] Christopher Hill, *God's Englishman*, p. 56. For Hill, Cromwell's skeptical advocacy of the free way is indebted primarily to the new science's critique of scholastic philosophy and Aristotelian science. By contrast, J.C. Davis, "Cromwell's Religion," in *Oliver Cromwell and the English Revolution*, pp. 181–208, argues that Cromwell's dedication to the free way is prompted less by science than by religion, specifically, Cromwell's providentialism and his anti-formalist Christianity.

[48] *The Writings and Speeches of Oliver Cromwell*, ed. Wilbur Cortez Abbott (Cambridge: Harvard University Press, 1937), I. 377.

and institutions as a hyper-masculine display of Machiavellian virtú.[49] But, the poem also surprisingly deploys the imagery of childbirth, nursing, and weaning to characterize Cromwell's free, category-bursting energies. Skeptical, anti-formalist concerns mix and merge with maternal matters most vividly when Marvell links the Lord General's purifying, fiery destructiveness and his innovative bursting and breaking energies with childbirth and nursing: "And, like the three-fork'd Lightning, first/ Breaking the Clouds were it was nurst,/Did through his own Side/His fiery way divide" (13–16).[50] In an unexpected intrusion of the birth room and nursery into the political sphere, Marvell domesticates Cromwell's epoch-changing energies by associating the relentless path of his "active star" with the violent bursting and breaking associated with childbirth and weaning. Cromwell, "like the three-forked Lightning," breaks through the storm clouds of war and regicide that "nurst" him.[51]

Marvell's Cromwell is both self-begotten and self-nurtured. He is mother and father to himself, an autonomous and self-sustaining source of power. Free from the influence of generations past, he "Urged his active Star" (l: 12). Marvell is deliberately ambiguous here. If Cromwell's decisive actions mark him as a modern, Machiavellian prince, his absolute and self-begotten autonomy also edges him dangerously close to the idealized image of the king as an androgyne (absolute in sexual terms) that James cultivates in *Basilikon Doron*. At the same time, however, Marvell also undermines these monarchical implications by emphasizing that Cromwell is as much a servant as he is a commander. Like "the three-fork'd Lightning" (l: 13), Cromwell's creative violence is a force of Mother Nature, and hence nourished by natural female powers larger than his own spectacular hyper-masculine display of personal autonomy and agency. In counterpoint with the *Ode*'s allusions to public affairs, both present and past, Marvell's allusion to birth and nursing helps to reconcile the impossible tensions between the paternal and the maternal, law and nature, destiny and personal agency, violence and emancipation, which the poet associates with

---

[49] Laura Lunger Knoppers, in *Constructing Cromwell: Ceremony, Portrait, and Print, 1645–1661* (Cambridge: Cambridge University Press, 2000), pp. 52–6, dwells less on the *Ode*'s Machiavellianism than on the ways in which the poem constructs a supremely masculine portrait of Cromwell as republican military hero.

[50] Andrew Marvell, *The Poems and Letters of Andrew Marvell*, ed. H.M. Margoliouth, 2 vols. (Oxford: Clarendon, 1971). Line numbers are cited in the text.

[51] Marvell's anti-monarchical image of childbirth suggestively contrasts with the "monstrous births" described in the "Mistress Parliament" satires penned by "Mercurius Melancholus," and published between 1648 and 1660. See Katherine Romack, "Monstrous Births and the Body Politic: Women's Political Writings and the Strange and Wonderful Travails of Mistris Parliament and Mris. Rump," in *Debating Gender in Early Modern England*, 1500–1700, pp. 216–19.

Cromwell's historically unprecedented and morally ambiguous career. Just as the maturing child inevitably rejects the breast that nursed it, so by the same natural law and cultural necessity must Cromwell pull away from "his own Side" (l: 15) so that he can single-mindedly fulfill his special destiny as "the Wars and Fortunes son" (l: 113). For Marvell, then, the language of nursing conveys the inevitability of sundering even the most intimate and natural bonds (between the nursing mother and child) in order to achieve a decisive break from kingship, to "cast the Kingdome old/Into another Mold" (ll: 35–6). As we shall see in Chapter 5, the same emphasis on breaking "the Bond of Nature" (*PL*, 9: 956) informs Milton's depiction of Samson as Israel's champion.

Cromwell also deploys the language of nurturing motherhood to activate new affective forms of social attachment capable of overriding the bonds of blood and natural kinship that determined rank, status, title, and other customary forms of social distinction and stature in "the Kingdome old" (l: 35). In several key parliamentary speeches, he draws on the language of maternal love and nurture to push past customs, titles, and other externals and to inspire a radically new sense of inward union, which alone is real. In September 1644, for example, Cromwell urges the Parliamentary committee appointed to convene with the Westminster Assembly, which generally favored a Presbyterian form of church government, to tolerate Independent and other separatist congregations who refused to conform to the Scottish model. Adopting the venerable biblical role of the nursing father, Cromwell maintains that the nurturing of the various sects is an essential part of his political mandate. "Endeavour the finding out some ways," he proclaimed, "how for tender consciences, who cannot in all things submit to common rule which shall be established, may be borne with, according to the Word, and as may stand with the public peace."[52]

The concerns of the Puritan household guides can be felt in Cromwell's expression of tender regard for "tender consciences." As we recall, tender mothers and tender children go together in guidebooks. The very word "tender" recurs in the guidebooks with a remarkable and highly suggestive frequency. Time and again, "tender"-ness is celebrated in the name of a new paradigm of family relations. Robert Cleaver glosses Proverbs 13.1, "A wise son hears his father's instructions," as meaning "a godly and prudent child of either sex" will accept rebuke only if rendered by "a most tender father." Citing David 4.3, he describes himself as "tender and only beloved in the sight of my mother."[53] Robert Pricke counsels fathers to be

---

[52] *The Writings and Speeches of Oliver Cromwell*, ed. W.C. Abbott, I., p. 294.
[53] Robert Cleaver and John Dod, *A Briefe Explanation*, 1615, pp. 207–8, 68.

"tender" with both their wives and children. Since a woman is "a weake and fraile vessell," Pricke writes, "therefore men are to deale with them in a tender and charie fashion: as men deale with glasses, and with tender vessels that are brittle." Children, writes Pricke, are "wholly addicted unto" the mother, because she "is most tenderly affected toward them."[54] In *Matrimoniall Honour*, Daniel Rogers advises that since a wife is "a thing naturally framed to tendernesse," a husband should be "tender" of her soul.[55] Oliver Heywood writes of his mother: "She was tenderly affected to the fruit of her wombe." In writing of his first wife, Elizabeth Angier, Heywood notes that, while her own mother died in 1640, her stepmother acted as a "very tender mother" in her stead.[56] William Gouge exhorts mothers to breast-feed their own babies by noting: "daily experience confirmeth" that maternally breast-fed children "prosper best. Mothers are most tender over them and cannot indure to let them lie crying out, without taking them up and stilling them, as nurses will let crie and crie again, if they be about any businesse of their owne."[57] Lady Macbeth's assertion that "I have given suck, and know/How tender 'tis to love the babe that milks me" (1: 7.54–5) before conveying willingness to dash out her baby's brains represents yet another, albeit unconventional, example of the "tender"-ness topos.

In presenting himself as a tender father who cares for the tender consciences of the dissenting sects, Cromwell operates squarely within the reformed paradigm of family/nation as an institution primarily dedicated to gently molding souls rather than forcing obedience. Speaking as the new nursing father, Cromwell aims at winning hearts and inspiring voluntary rather than conscripted service. By creating affective bonds between himself and his soldiers and inspiring powerful feelings of community within the Army, Cromwell hoped that a felt sense of solidarity would override the points of doctrinal contention that threatened to divide the Army. By the same means, he also tried unsuccessfully to temper the increasingly adversarial relations between the Army and Parliament. At the Army Council called on July 16–17, 1647, the Agitators (the delegates that the Army elected as its representatives) moved to march directly against Parliament, which had repeatedly reneged on its promise to meet the Army's demands for financial restitution. Cromwell urged that the Army should proceed first by persuasion not force, and that

---

[54] Robert Pricke, *The Doctrine of Superiority*, Section K.

[55] Daniel Rogers, *Matrimoniall Honour: or The mutuall Crowne and comfort of godly, loyall, and chaste Marriage*... (London, 1642), p. 242.

[56] *Diaries of Oliver Heywood*, ed. J.H. Turner (Brighouse, England, 1883), I., pp. 50–4, 58.

[57] William Gouge, *Of Domesticall Duties*, p. 289.

rather than march on Parliament it should work instead to strengthen the influence of the minority of the Commons sympathetic to the Army's cause: "that [which] we and they gain in a free way, it is better than twice so much in a forced, and will be more truly ours and our posterity…Though you be in the right and I in the wrong, if we be divided I doubt we shall all be in the wrong."[58]

In taking the free way to union, Cromwell assumes the same mode of tolerant, nurturing motherhood that governs reformed seventeenth-century rhetoric of family life. To achieve a unity that is "truly ours" within the Army and between the Army and Parliament, Cromwell attempts to engage the hearts and nurture the spiritual commonalities within the Army and between the Army and Parliament, lest superficial differences (whether "you be in the right and I in the wrong") splinter the substantial whole. The bonds of familial affection that Cromwell attempts to cultivate through his political rhetoric of paternal–maternal nurture close the distance between the speaker and his various audiences. Despite manifest class differences and doctrinal disagreements, "[Y]ou" and "I" are on the same footing, and both are equally implicated in the potential break-up of the "all": "if we be divided I doubt we shall all be in the wrong."

The same reformed image of Cromwell as nurturing father surfaces in motivated ways in regicidal and republican accounts of the October 1641 Irish Rebellion, especially in contrast to Charles's hard-heartedness and lack of compassion for his own people, as already noted in the previous section. In *The Tenure of Kings and Magistrates* (1649), Milton depicts a hardened King Charles. The king and his Presbyterian allies, writes Milton, had caused "so great a deluge of innocent blood" that they "cannot with all thir shifting and relapsing, wash off the guiltiness from thir own hands" (*CPW*, 3: 197, 227).[59] David Loewenstein observes in Milton's image of the guilty king with innocent blood on his hands an interesting Shakespearean subtext at work in Milton's tract. For Loewenstein, Milton links "the prevaricating king and the Scottish Presbyterian clergy supporting him by recalling the blood guilt of Lady Macbeth."[60] Following up on this suggestive possibility, I would add that by comparing Charles to Lady Macbeth, the literary exemplar par excellence of maternal viciousness, Milton also underscores Charles's unmanly lack of paternal compassion, his dried-up milk of human kindness, and thus the hollow, superficial

---

[58] *The Writings and Speeches of Oliver Cromwell*, ed. W. C. Abbott, I. pp. 483, 486.

[59] All references to Milton's prose are taken from *The Complete Prose Works of John Milton*, ed. Don M. Wolfe et al., 8 vols. (New Haven and London: Yale University Press, 1953–82), and are noted in the text.

[60] David Loewenstein, "The King Among the Radicals," in *The Royal Image*, p. 106.

nature of Stuart kingship, its status as empty icon. This comparison gestures as well toward Cromwell, who by contrast to hard-hearted Charles, emerges as the true nursing father. Milton thus steals back this venerable biblical trope from the Stuarts and aligns it instead with the anti-dynastic and regicidal concerns of his tract.

The troubled history of Anglo-Irish relations, however, complicates the Cromwellian figure of the nursing father.[61] As Milton's politically motivated use of the trope against Charles and his alliance with the Irish Rebels reveals, Cromwell's discourse of nurture not only exposes the king's blood guilt, but it also fosters what Paul Stevens terms "Leviticus thinking": the radical Protestant exclusionism that finds scriptural justification for imperial conquest (i.e. Cromwell's brutal campaign against Ireland in 1650), by "reviving many of the [Mosaic] Law's intensely ethnocentric, community-identity imperatives."[62] In *Observations upon the Articles of Peace with the Irish Rebels*, commissioned by Cromwell's parliament on March 28, 1649, Milton reflects the views of the Cromwellian regime when he speaks of "*Irish* Barbarians" (*CPW*, 3: 308) and of "the villainous and savage scum of Ireland" (*CPW*, 4: 323) to justify Cromwell's "civilizing Conquest" (*CPW*, 303). By deploying the biblical trope of the nursing father, Cromwell legitimates his campaign for religious tolerance at home *and* his imperial aggression abroad. Responding partly to republican fears of an Irish-based Stuart invasion, partly to English fantasies of Ireland's participation in a worldwide, Counter-Reformation conspiracy, and partly to the Protestant exclusionism that renders the Irish as the barbarous Other, Cromwell embarks on a violent campaign of ethnic cleansing in Ireland in 1650.

These brutal acts of imperial conquest might seem at odds with Cromwell's carefully cultivated image as the nursing father who tolerates sectarian differences. But, when read as a reflection of the emergent nation's foundational paradoxes, Cromwell's religious toleration and his imperial aggression in fact acquire ideological coherence. Although already noted, it is important to reiterate in this context that the modern nation is Janus-faced: humanitarian and bellicose, bounded and expansionist, tolerant at home and intolerant abroad, grounded in reason and law, yet capable of arousing passionate feelings of patriotism, which, in turn, can

---

[61] Willy Maley cautions against reducing the tensions between Ireland and England to a simple binary opposition in "Milton and 'the complication of interests' in Early Modern Ireland," *Milton and the Imperial Vision*, ed. Balachandra Rajan and Elizabeth Sauer (Pittsburgh: Duquesne University Press, 1999), p. 157.

[62] Paul Stevens, "Milton's Janus-faced Nationalism": 259. See also "'Leviticus Thinking' and the Rhetoric of Early Modern Colonialism," *Criticism*, 35 (1993): 441–61, and "Spenser and Milton on Ireland: Civility, Exclusion, and the Politics of Wisdom," *Ariel*, 26 (1995): 151–67.

justify brutal acts of imperial aggression and colonial occupation.[63] On the one hand, the modern nation unifies unlike peoples of diverse faiths, races, classes, and regions by inspiring a passionate sense of voluntary cooperation and patriotic sacrifice capable of overriding all local allegiances. On the other, this new tolerant sense of national unity and unforced allegiance to the nation also justifies state-sponsored aggression toward others abroad.[64] In making these incompatible concerns cohere at his nurturing breast, Cromwell, as nursing father, for better or worse, plays a central role in shaping England into a modern nation. Christopher Hill observes that:

> The British Empire, the colonial wars which built it up, the slave trade based on Oliver's conquest of Jamaica, the plunder of India resulting from his restitution and backing of the East India Company, the exploitation of Ireland; a free market...religious toleration, the non-conformist conscience, relative freedom of the press,...none of these would have come about in quite the same way without the English Revolution, without Oliver Cromwell.[65]

As Cromwell's speeches vividly demonstrate, the language of maternal love and nurture grants rhetorical and affective power to the emergent modern nation's foundational contradictions.

These paradoxes become even more pronounced in the shift from Cromwell's Commonwealth to his Protectoral governments. There is a correlative shift in the Lord Protector's rhetoric of nurture. Just as he had advocated in 1644 for tender consciences, Cromwell tries again in 1653 to remedy the deepening rifts between competing religious and political constituencies through maternal discourse. Adopting the role of the mother-minister, he urges the Barebones Parliament to shore up the national whole by embracing and nurturing all of its parts: "Therefore I beseech you have a care of the whole flock! Love the sheep, love the lambs; love all, tender all."[66] Although in this instance, the term "tender" is deployed as an imperative rather than as an adjective, which is by far the more customary form, this last heartfelt command to "love all, tender all" can be read as a variant example of the "tender"-ness *topos* familiar from the guidebooks. It highlights once again how Cromwellian appropriations of the maternal

---

[63] See Linda Gregerson, "Colonials Write the Nation: Spenser, Milton, and England on the Margins," in *Milton and the Imperial Vision*, ed. Balchandra Rajan and Elizabeth Sauer (Pittsburgh: Duquesne University Press, 1999), p. 169, and Elizabeth Sauer, "Religious Toleration and Imperial Intolerance," *Milton and the Imperial Vision*, pp. 214–17.

[64] The "Janus-faced nation" is Tom Nairn's term. Tom Nairn, *The Break-up of Britain: Crisis and Neo-Nationalism*, 2nd ed. (London: Verso, 1981).

[65] Christopher, *God's Englishman: Oliver Cromwell and the English Revolution*.

[66] Abbott, ed., *Writings and Speeches*, III., p. 62.

aim at affectively uniting the divided social whole. Through love, tenderness, and fellow feeling, the transcendent "all" invoked in the speech will be made manifest.

But while the maternal rhetoric of this Protectoral speech echoes that of the speeches made in the 1640s, the historical context is quite different. On April 20, 1653 Cromwell shut down Parliament by force; in December of that same year, he was installed with great fanfare and ceremonial flourish as England's Lord Protector. Cromwell was given the title of king in early drafts of the Instrument of Government, England's first written constitution. Although he himself declined the throne, Cromwell was depicted as a would-be king in a variety of media, from the new coinage to the revival of Augustan poetry.[67] Cromwell did little himself to counter this image. Reviving a customary monarchical practice, he made a formal entry to London on February 8, 1654. Like the Stuart kings before him, he asserted his personal rule when, in February 1655, he dissolved his first Protectoral Parliament and shut down the press, except for two official newsbooks. Poets like Andrew Marvell in *The First Anniversary of the Government under His Highness the Lord Protector* brilliantly argued for the marked differences between Protectoral and monarchical government, but they too criticized Cromwell, "*His Highness*," for his monarchical inclinations.[68]

The language of maternal love and nurture in Cromwell's Protectorate speeches thus reflects the ambiguities inherent in this complex epoch-shifting moment. On the one hand, Cromwell's appeal to a unity achieved through maternal love and tenderness hearkens back to his pre-Protectoral emphasis on fostering a nurturing spirit of Union that would override the points of doctrinal differences that thwarted collective action and impeded Reformation. The continuities between this speech and the earlier speeches also encourage a perception of similar continuities between the

---

[67] For illuminating analysis of the contested image of Cromwell in the early and late Protectorate, see Laura Lunger Knoppers' excellent study, *Constructing Cromwell: Ceremony, Portrait, and Print, 1645–61* (Cambridge: Cambridge University Press, 2000), Chapters 3 and 4. See also David Norbrook, "King Oliver? Protectoral Augustanism and its Critics, 1653–58," in *Writing the English Republic*, Chapter 7.

[68] Excellent studies that focus on the tensions between Cromwell's monarchical and Protectoral image in *The First Anniversary* include Knoppers, *Constructing Cromwell*, pp. 98–102; Norbrook, *Writing the English Republic*, pp. 339–51; J.A. Mazzeo, "Cromwell as Davidic King," in his *Renaissance and Seventeenth-Century Studies* (New York: Columbia University Press, 1964), pp. 183–208; Steven Zwicker, "Models of Governance in Marvell's 'The First Anniversary'," *Criticism* 16 (1974): 1–12; M.L. Donnelly, "'And still new stopps to various time apply'd': Marvell, Cromwell, and the Problem of Representation at Mid-century," in *On the Celebrated and Neglected Poems of Andrew Marvell*, ed. Claude J. Summers and Ted-Larry Pebworth (Columbia: University of Missouri Press, 1988), pp. 163–76; Annabel M. Patterson, *Marvell and the Civic Crown* (Princeton: Princeton University Press, 1978).

Commonwealth and the Protectorate. Toleration provides the connecting thread. As in the just-cited 1653 speech, Cromwell hopes to unite the "whole flock" by dispensing love and tenderness to all the dissenting sects. In exhorting Parliament to "love all, tender all," Cromwell tailors the allegory of the Church's love for the penitent sinner from Canticles to suit his political interest in promoting toleration. Just as John Cotton writes that ministers should "apply themselves to the estate of their people: If they bee babes in Christ, to be as breasts of Milke to suckle them,"[69] so Cromwell urges the MPs to follow his lead in tolerating the Independent and Nonconformist churches.

On the other hand, however, Cromwell's rhetoric of maternal nurture also highlights the new monarchical aspects of his Protectoral image. Although his exhortations to love and tender reflect his emphasis on religious toleration, the same rhetoric of mother-love also reflects expansionist ambitions similar to those shaping James's self-image as a "a louing nourish-father" *(BD*, 24). Imperial ambition, expanded personal rule, and the campaign for spiritual bonds of fraternal affection intertwine, somewhat incoherently, in Cromwell's Protectoral rhetoric.[70] James's campaign for Union between Scotland and England goes hand in hand with his efforts to conquer the hearts of his subjects. Not dissimilarly, Cromwell's Protectorship is governed by the seemingly impossible overlap between his imperial expansionism and religious tolerance, as in the simultaneity between his ill-fated Western Design and his proposal in 1655 for Jewish Readmission. Having failed to unite the divided nation by unsuccessfully waging war against Spain in the West Indies, Cromwell intensified his campaign to achieve national union through toleration by defending the limit-case of this policy: the readmission of the Jews. (Cromwell's support for toleration proved no more successful than his acts of imperial aggression. By early 1656, the proposal for Jewish Readmission had failed.) When read in light of the aggressive, imperial stance that he takes toward achieving national union in both his war adventures abroad and his all-out campaign for toleration at home, the intense pressure that Cromwell puts on the need to love all acquires different nuances and a heightened sense of urgency. In his desperation to achieve union, Cromwell pulls out all the stops. All means, love and war, maternal nurture and paternal aggression, must be simultaneously deployed so that national union could be achieved.

[69] John Cotton, *A Brief Exposition of the whole Book of Canticles* (London, 1642), pp. 240–1.
[70] Elizabeth Sauer, "Religious Toleration and Imperial Intolerance," in *Milton and the Imperial Vision*, ed. Balachandra Rajan and Elizabeth Sauer (Pittsburgh: Duquesne University Press, 1999), p. 217.

Thus, whereas in the Commonwealth years, Cromwell, as a nursing father, operates inside the new paradigm of the family in his earlier speeches, his nurturing paternal image as Protector edges closer to a more traditional model of personal rule and patronage, as Cromwell's republican critics charged. Norbrook observes that "It is perhaps difficult today to reconstruct what an affront the title of 'Protector' would have been to [Cromwell's republican critics'] central principles, which implied that the people were incapable of looking after themselves without a supremely enlightened superior."[71] A correlative emphasis on self-sufficiency can be felt in reformed writings on the new family. *The Child-Bearers Cabinet* (1652) advises that children between the ages of three and seven "are to be educated gently and kindly, not to be severely reprehended, chidden, or beaten, for that means they may be made throughout their whole life after too timorous, or too terrified, astonished, and sotted."[72] Rather than force children into submission and encourage their subservience and docile embrace of authority, nurturing parents were to foster independent thought and feeling, or liberty of conscience, in their children. When a child reached maturity, he would have achieved the self-sufficiency and personal agency required to function in the modern post-dynastic state and in an increasingly entrepreneurial marketplace. The mature citizen-subject requires neither parental or governmental supervision nor coercion. Liberty of conscience both grants ethical self-direction and inspires voluntary service to the nation state. As Lord Protector, Cromwell in the 1650s moves closer to the imperial image of "louing nourish-father"-hood that James cultivates in *Basilikon Doron* (24). He does not, however, completely give up on the reformed family/nation's interlocking principles of nurture and liberty, to which he sustained his commitment by advocating religious toleration. As the Cockermouth annalist proclaims, Cromwell is "the nursing father of the churches."[73]

Cromwell's language of maternal love and nurture helps to highlight the shifting political fault lines in his Protectoral image. As Knoppers argues, "unlike the unity of monarchical portraiture...images of Cromwell were varied, shifting, and contested."[74] In part, such contestations result from the passive role that Cromwell took in shaping his public persona; they also reflect the political complexities of this historically specific

---

[71] David Norbrook, *Writing the English Republic*, p. 319.

[72] *The Child-Bearers Cabinet* (London, 1652), cited by Joseph E. Illick, "Child-Rearing in Seventeenth-Century America," in *The History of Childhood*, ed. Lloyd de Mause (New York, 1974). See also Leverenz, p. 75.

[73] Quoted in Claire Cross, "The Church in England 1646–1660," in *The Interregnum: The Quest for Settlement*, ed. G.E. Aylmer (Hamden, CT: Archon-Shoe String, 1972), p. 117.

[74] Knoppers, *Constructing Cromwell*, p. 80.

interval. On the one hand, these contestations parallel the unraveling of the Cromwellian regime. On the other hand, however, the disparities shaping Cromwell's image as nursing father help to consolidate his power, enabling him simultaneously to address widely different political constituencies, monarchist and separatist, while maintaining a sense of continuity despite the radical shifts and turns in his domestic and foreign policies and the changing nature of his own political role. If during the Protectorate years, Cromwellian appropriations of maternal nurture make it possible for the coercive power of the state to be articulated as paternal–maternal tenderness, in the Commonwealth era, the same rhetoric helps to encourage a love of mother country expansive and flexible enough to unite a diverse citizenry into a unified national whole.

"HAPPY NURTURE" IN *OF EDUCATION*

Basilikon Doron stages the ideal father–son relationship as a scene of instruction. James proclaims his monarchical authority through a book whose lessons his son and other readers must master in order to receive his paternal blessing. The "Praeceptor," writes James, must be honored as one of "them that are *in loco Parentum* vnto you" (*BD*, 41). Milton's *Of Education* offers a republican response to this monarchical equation among family, school, and kingdom—a response made all the more urgent by the fact that the universities were firmly under Royalist control when *Of Education* was published anonymously in 1644. The tract represents an attempt to wrest control of the training and disciplining of young minds away from the Crown. Milton aligns the new academy with the needs of the new state by designing a curriculum that would replace "empty rhetoric" with "ripe wisdom," and "servitude to fashion and flattery" with "responsible action." Whereas the universities fashion fawning courtiers and frightened counselors and clerics, "with souls so unprincipl'd in virtue, and true generous breeding," Milton's reformed course of study would provide the new nation with strong-minded and principled statesmen capable of leading England into the reformed future. Pupils should be taught to "know the beginning, end, and reasons of political societies; that they may not in a dangerous fit of the commonwealth be such poor, shaken uncertain reeds, of such tottering conscience, as many of our great counsellars have lately shewn themselves, but stedfast pillars of the State" (*CPW*, 2: 375).

The first of Milton's tracts to be licensed and registered, *Of Education* was written as a letter to the educational reformer Samuel Hartlib, who repudiated rote learning, and whose influence can be felt in Milton's attacks on the universities. Hartlib was a follower of the Moravian minister John Amos Comenius, perhaps the most influential Puritan educational reformer of the time. Comenius had been Hartlib's guest during his 1640–41 stay in England. As readers have noted, the influence of Comenius's utilitarian and empirical biases can be felt in Milton's discussion of curriculum. Juan Luis Vives's *De tradendis desciplinis* (1531) is another well-known source for Milton's tract, as are Robert Ascham's *The Schoolmaster*, Thomas Mulcaster's *Positions*, and other texts representative of the humanistic tradition.

Surprisingly scant attention, however, has been paid to the influence of popular domestic manuals on nursing such as Cleaver and Dod's *A Godly Form of Household Gouernement* and William Gouge's oft-reprinted *Of Domesticall Duties*, despite the fact that Milton himself indirectly alludes to such texts at the end of his tract.[75] Milton writes that if "brevity had not been my scope," he might have addressed the care and teaching of infants and young children, "beginning, as some have done, from the cradle, which yet might be worth many considerations" (*CPW*, 2: 414–5). Milton may also have Erasmus's colloquy *Puerpa* (1526) in mind. Like the guidebook writers, Erasmus exhorts mothers to nurse their own children, while pointing up the close connections between household and political governance. Vivid examples of anarchy, court bankruptcy, peasant revolt, and religious dissent in Denmark, France, and Germany, and a general sense of widespread social unrest frame Eutapelus's insistence that the sixteen-year-old Fabulla breast-feed her own baby. As Barbara Correll argues, Erasmus both places the nursing mother as an "icon of organic wholeness" in sharp contrast to the riot and disorder outside the home and sees her as a "symbolic restoration of order in the larger sphere."[76] Milton clearly finds such matters "worthy [of] many considerations." Whether it is the guidebooks writers or Erasmus, or both, the general influence of those who address the nursing, nurturing, and teaching of children, "as some have done from the cradle" (2: 414–5), finds expression in *Of Education*, most specifically, in the detailed attention that Milton pays to the nourishment of students' minds and bodies. For Milton, "happy nurture" (*CPW*, 2: 377) is crucial to the shaping of character, the building of strong bodies, and the proper schooling of the soul.

Just as the new nurturing family is a crucial site for the reformation of the national subject, so too is Milton's reformed classroom. *Of Education*'s

---

[75] Milton also cites classical authorities that treat early childhood in their educational programs, including Plato (*Laws*, I, 643bc, and VII, 793e) and Quintillian (*Institutio Oratorio*, I, I, 3–5). See *CPW*, 2. 414, n. 38.

[76] Barbara Correll, "Malleable Material, Models of Power: Women in Erasmus's 'Marriage Group' and *Civility in Boys*," *ELH*, 57 (1990): 241–62.

innovative consolidation of the conflicting moral and empirical-utilitarian emphases in sixteenth- and seventeenth-century educational philosophy provides a distinctively modern program for inserting the reformed individual into the new nation.[77] In Milton's tract, induction provides the connecting thread between conflicting moral and utilitarian agendas of reformed education and its national application. Most relevant here is that it also provides an optimal means for achieving reformation in affective terms: for changing servile, tradition-bound students, fearful of challenging the ancients, into intrepid, forward-looking seekers of universal principles. As already cited, *The Child-Bearers Cabinet* advises that children between the ages of three and seven "are to be educated gently and kindly, not to be severely reprehended, chidden, or beaten, for that means they may be made throughout their whole life after too timorous, or too terrified, astonished, and sotted."[78] Not dissimilarly, Milton argues that a gentle and gradual, inductive approach to the acquisition of knowledge should replace the stern, forbidding schoolmaster, who frightens or forces his students into submission. The rhetoric of nurture is crucial to his anti-customary, activist, and inductive way to acquire the "ripest judgment" (*CPW*, 2: 374) necessary to create patriotic English citizens.

For Milton, proper learning must "begi[n] with arts most easie, and those be such as are most obvious to sence." Under the old system, students grapple prematurely with "the most intellective abstractions of Logic & metaphysics" without sufficient time to work through more embodied modes of thinking. "These [abstractions] are not matters to be wrung from poor striplings," Milton maintains. "The empty wits of children" should not be compelled to perform "the acts of ripest judgment." The universities fail to distinguish children from adults, "empty wits" from "ripest judgment" (*CPW*, 2: 374, 372). For Milton, "the right path of a vertuous and noble Education" takes students from childhood to adulthood on into citizenship (*CPW*, 2: 376). Domestic and educational reform collides in the tract: nurture not force is the mainspring of the new education. Like the guidebook writers, Milton advocates nurture as the best means for molding children's hearts and minds and for preparing them to become judicious and self-directed adults, "fittest to chuse," and thus capable of active participation in the life of the nation. While "our choisest and hopefullest wits" must now be forcibly dragged to "that

[77] Gauri Viswanathan, "Milton and Education," in *Milton and the Imperial Vision*, ed. Balachandra Rajan and Elizabeth Sauer (Pittsburgh: Duquesne University Press, 1999), pp. 273–93.

[78] *The Child-Bearers Cabinet* (London, 1652), cited by Joseph E. Illick, "Child-Rearing in Seventeenth-Century America," in *The History of Childhood*, ed. Lloyd de Mause (New York, 1974). See also Leverenz, p. 75.

asinine feast of sowthistles and brambles which is commonly set before them," even "our dullest and laziest youth" will be inspired by "the infinite desire of such a happy nurture" perfectly suited to "their tenderest and most docible age" (*CPW*, 2: 376–77).

Not unlike the natural nursing mother in the guidebook literature, Milton's new academy would provide those of a "tenderest and most docile age" with proper educational nourishment (2: 377). Tender schools and tender students go together. With a delectable but age-appropriate educational diet, learning is associated with the cultivation of moral character and physical well-being. This gentle process of character-formation and healthy eating and exercise culminates organically in the expression of patriotism—the love of mother country. By contrast, students, under the present system, care neither for family nor country. Rather than gaining "worthy and delightful knowledge," pupils have been "mocked and deluded all this while with ragged notions and babblements." A poor educational diet results in disabled and disaffected learners, weaned too soon from "sensible things." Improperly cared for and nourished, students grow into emotionally and intellectually stunted lawmakers, judges, and statesmen, who both cannot and care not to distinguish "the prudent and heavenly contemplation of justice and equity, which was never taught them" from "the promising and pleasing thoughts of litigious terms, fat contentions and flowing fees." These are men with "souls so unprincipled in vertue and true generous breeding, that flattery and court shifts and tyrannous aphorisms appear to them the highest points of wisdom" (*CPW*, 2: 373–4).

As poor nurturers, the universities thus replicate the role of our first parents, to return to the assertion with which the tract begins. Just as Adam and Eve brought death into the world by succumbing to an unwholesome appetite for forbidden knowledge, which they transmitted to their innocent offspring, so too do the universities replay humankind's fall into moral and physical degeneration by feeding the most tender students a harsh, non-nurturing diet of imperfect and perverse knowledge, corrupting both their minds and hearts. By way of restoring the mind and heart to their original condition, Milton's tract ends by briefly considering where, when, and what students should eat as matters crucial to the cultivation of a moral character. Just as lute or organ music helps "to smooth and make [students] gentle from rustick harshnesse and distemper'd passions," so such music would "not be unexpedient after meat to assist and cherish nature in her first concoction." After exercising, "a convenient rest before meat" is suggested (*CPW*, 2: 410–11). With much greater invective, Milton makes a similar point about the intimate relations between diet and moral character in the anti-prelatical pamphlets of 1641–42, when he characterizes the Anglican clergy as "fat and fleshy" (*The Reason of Church Government*, 1, 858), degenerate worshipers of "*Mammon* and Their Belly" (*Of Reformation*, 1: 566), and as possessing "many benefice-gapping mouth[s]" and "canary-sucking and swan-eating palate[s]" (*Of Reformation*, 1: 549). Milton's discussion of diet at the end of *Of Education* drives home this same linkage among "happy nurture" (2: 377) and the making of moral subjects.

Just as kindly educational nurture allows both the body and mind to flourish, so it also gently disciplines the affections. Milton underscores that every stage of the intellectually redemptive ascent from "sensible things" to "things invisible" provides an occasion for cultivating pupils' affections and pointing them toward a social productive end. "But here the main skill and groundwork," Milton writes, "will be to temper them such lectures and explanations upon every opportunity, as may lead and draw them in willing obedience, enflam'd with the study of learning and the admiration of vertue, stirr'd up with high hopes of living to be brave men and worthy patriots, dear to God and famous to all ages." With "mild and effectuall perswasions," teachers will gently "infus[e] into their young brests such an ingenuous and noble ardor, as would not fail to make many of them renowned and matchlesse men" (*CPW*, 2: 384–5). Under the old system, students remain "childish," "ill-taught," and ill willed, and intellectually and emotionally undisciplined. In the new nurturing academy, by contrast, they ultimately shall possess a will that is both good and free, gain "incredible diligence and courage," and break free from carnal, servile thinking. At this highest point of affective and intellectual development, childhood will come to an end: students will "despise and scorn all their childish and ill taught qualities to delight in manly, and liberall exercises." Maturity also marks the birth of a national consciousness. Milton's students enter school as boys and, nine years later, filled with "happy nurture," they emerge as politically enlightened men, poised to leave the academy and enter public life as "brave men and worthy patriots," "manly" citizens (*CPW*, 2: 385).

NURSING LIBERTY IN *AREOPAGITICA*

In *Areopagitica*, Milton elaborates on many of the same concerns about nurture, character building, and curriculum that he discusses in *Of Education*. In both tracts, which were published in the same year of 1644, the free way to Truth, the emancipation of the intellect, and the making of moral subjects represent interlocking concerns. Like the ideal graduate of Milton's new academy, the ideal citizen imagined in *Areopagitica* is marked by his capacity for mature thought and feeling. Milton has both the un-formed schoolboy and the mature male citizen in mind when he critiques Parliament for passing the Licensing Order of June 14, 1643, restoring the system of pre-publication censorship, which had collapsed after the outbreak of civil war. "What advantage," he asks, "is it to be a man over it is to be a boy at school, if we have only scapt the ferular, to come under the fescu of an *Imprimatur?*" (*CPW*, 2: 531). Just as the universities force "the empty wits of children" to perform "the acts of ripest judgment" (*CPW*, 2: 374), so Parliament would "captivat" mature citizens "under a perpetuall childhood of prescription" (*CPW*, 2: 531).

It is the popish and patriarchal character of the Licensing Order to which Milton particularly objects: he repudiates not only the Order but all forms of "tolerated Popery and open superstition" as well. In an argu-ment calculated to exacerbate mounting fears that the King had formed an alliance with international Counter-Reformation forces, Milton main-tains that, by restoring pre-publication licensing, Parliament had played into the hands of the Pope by revitalizing the aims of both the Council of Trent and the Spanish Inquisition, which "engendering together brought forth, or perfected those Catalogues, and expurgating Indexes that rake through the entrails of many an old good Author with a violation worse than any could be offered to his tomb." By imposing popish control over the minds and hearts of the English people, Parliament had made "us now lesse capable, lesse knowing, lesse eagerly pursing of the truth" (*CPW*, 2: 559). As his trice-repeated "lesse" underscores, Milton emphasizes that Parliament's assertion of papal-like authority had diminished and de-pressed the English people, who had only just begun to assert their reli-gious and political independence and to experience the joy that comes with the experience of such freedoms. Rather than point the present moment toward a salvific future, Parliament had turned the epoch-changing events of the early 1640s back toward the dark, outmoded Catholic past.

Milton presents his case against the Licensing Order in the form of a classical oration, whose title acknowledges Isocartes's "Areopagiate Dis-course" as its chief influence. The tract's title-page quotation from *The Sup-pliant Women* also evokes the "spirit of bold thinking and speaking" epitomized by Euripides's plays.[79] As John Steadman has documented, Ar-istotle and Quintillian serve as models for *Areopagitica*'s rhetorical pattern. Milton's tract can also be set squarely within the humanistic tradition.[80]

[79] David Norbrook, *Writing the Republic*, p. 127.
[80] John M. Steadman, "The Dialectics of Temptation: Milton and the Idealistic View of Rhetoric," in *The Hill and the Labyrinth: Discourse and Certainty in Milton and His Near-Contemporaries* (Berkeley: University of California Press, 1984), p. 71.

Very little attention, however, has been paid to the maternal imagery that conspicuously surfaces in the text. Even the most recent Milton editions gloss over the importance of such images, despite that, as Katherine Maus observes, Milton (like Sidney, Jonson, Shakespeare, and many other English Renaissance poets) "relies in *Areopagitica* upon extended analogies between 'the issue of the brain' and the 'issue of the womb'."[81] Time and again, Milton mixes in maternal and domestic metaphors with his political arguments and invective. For example, he writes that, before the Council of Trent and the Spanish Inquisition "engendering together brought forth" the papal Catalogues and Index, "Books were ever as freely admitted into the World as any other birth...no envious *Juno* sate cross-leg'd over the nativity of any mans intellectuall off-spring." Elsewhere in the tract, Milton repudiates the Licensing Order as "a step-dame to Truth," even as, despite itself, the renewal of pre-publication censorship "may prove a nursing mother to sects" (*CPW*, 2: 543).

Notably, *Areopagitica*'s resonant passages on liberty contain the poet's most compelling evocations of maternal nurture and nursing. Milton defines liberty as "the nurse of all great wits." Paralleling the nurture of liberty with "the influence of heav'n," Milton proclaims that "this is that which hath rarify'd and enlighten'd our spirits...this is that which hath enfranchis'd, enlarg'd and lifted up our apprehensions degrees above themselves" (*CPW*, 2: 559). Milton's image of liberty as the nurse of uplifted and enlightened spirits explicitly recalls Longinus's classical formulation of the sublime in *Peri Hypsous*: "a *Democracie* is the best *Nurse* of high Spirits...just liberty *feeds* and *nourishes* the thoughts with great *notions*, and *draws* them forward."[82] Milton's Longinian image of liberty as the nurse and nourisher of the sublime shapes his description of London as "the mansion house of liberty," the tract's most passionate nationalist image. In a pile-up of over-full and overflowing sentences, Milton demonstrates how liberty in London nurses the innovative, category-bursting, intellectual energies of the sublime: "then there be pens and heads there, sitting by their studious lamps, musing, searching, revolving new notions and idea's wherewith to present, as with their homage and their fealty the approaching Reformation: others as fast reading, trying all things, assenting to the force of reason and convincement." Milton builds towards a crescendo of activity, "musing, searching, revolving...trying, [and] assenting," as he tries to capture the profusion and swift pace of "new notion and idea's" when the mind is nursed by liberty (*CPW*, 2: 554).

[81] Katharine Eisaman Maus, *Inwardness and the Theater in the English Renaissance* (Chicago: University of Chicago Press, 1995), p. 185.

[82] Longinus, *Peri Hypsous, or On the Sublime*, pp. 78–9.

Milton's portrait of reformed leadership closely reflects the governing ideals not only of his new school but also of the new family, with its emphasis on nurturing souls, nourishing character, and gently cultivating strong minds and hearts rather than forcing bodies and consciences into submission. He applauds Parliament for its "milde and equall Government," "meek demeanour," "gentle greatnesse," "magnanimity," and "happy counsels." To these kindnesses he opposes the "jealous haughtinesse of Prelates and cabin Counsellours," "inquisiturient bishops," and "the laziness of a licensing church" (*CPW*, 2: 488–9, 559). In passing its ordinance, a once-gentle and nurturing Parliament had come to resemble the horribly punitive kings and magistrates it had hitherto opposed. Rather than cast off the paternal constraints and inquisitorial severity characteristic of the old family, Parliament had instead revived and reinvigorated these outmoded forms of patriarchal and papal authority. Since God himself trusts man "with the gift of reason to be his own chooser," Parliament's thwarting of the English people's intellectual and moral independence by reimposing pre-publication censorship not only slows the rapid pace and intellectual generativity of Reformation, but it also denies the divine will. Rather than keep "the censor's hand" on the backs of the English people and "a slavish print upon our necks," Parliament must resist "the gripe of custom" and show "som grain of charity" (*CPW*, 2: 564, 554). It is not accidental that charity is central to his model of new government. Although he does not explicitly allegorize charity as a nursing mother, Milton's depiction of charity evokes mother-love, compassion, and nurture: "all charitable and compassionate means be us'd to win and regain the weak and misled" (*CPW*, 2: 565). Wresting charity (and its association with the nursing mother) away from the Stuarts and making it serve the cause of liberty and Reformation instead, Milton emphasizes that a new emphasis on toleration and the love of God and neighbor must prevail over "the censor's hand" and "the gripe of custom" (2: 564, 554).

Unlike *Of Education*, *Areopagitica* explores the relations between nurture and nation within the distinctly Pauline framework that organizes Milton's social vision especially in his late poetry, most specifically, in the last two books of *Paradise Lost*. As Jason Rosenblatt argues, Milton's Hebraism, his "living appreciation of the spirit and content of Torah," complicates his Paulinism. Especially between 1643 and 1645, when *Areopagitica* was published, "Milton felt and recorded what Paul never did, the saving power of the Pentateuchal law."[83] An accomplished Hebraist, Milton, in *Of Education*, recommends that students learn Hebrew

---

[83] Jason P. Rosenblatt, *Torah and Law in Paradise Lost* (Princeton: Princeton University Press, 1994), p. 64.

so "that the Scriptures may be now read in their own originall." Milton's ambivalence toward the Hebraic finds particularly clear expression in *Areopagitica*, when Milton addresses the roles that nurture and loving kindness play in the making of moral subjects. Is the Law nurturing and life-sustaining, as rabbinic commentary proclaims, or is it a death sentence, as Pauline doctrine asserts (Rom. 8–2–3, Gal. 3:21)? Deploying the topos of maternal tenderness, John Lyly in *Euphues* describes God as a nurturing father who molds a divided England into a unified new Israel: "So tender a care hath [God] alwaies had of that *England,* as of a new *Israel,* his chosen and peculiar people."[84] Like Lyly, Milton in *Areopagitica* depicts the English people as the children of a tender, nursing deity, but he remains ambivalent about whether or not God's divine nurture encompasses both Hebraic law and Christian charity.

As Rosenblatt suggests, the Amsterdam rabbi Menasseh ben Israel's 1639 Latin work, *De Termino Vitae,* may be a source for one of Milton's outlines for a tragedy on the theme of "Paradise Lost" contained in the Trinity Manuscript.[85] Menasseh's scientific and miraculous explanations of the Law's life-giving and life-sustaining properties prompt comparison with the emphasis in the sermon and guidebook literature on gospel as mothers' milk—texts which as we have seen Milton indirectly alludes to at the end of *Of Education.* Menasseh writes that "Our Life may be prolong'd by observing God's Laws," because the Law miraculously preserves the "radical Moisture" and "the Vertue of our Food," as is "confirmed by the Example of *Moses,* who liv'd forty Days in Mount Sinai, without Meat or Drink; then the Divine Law (which was instead of Food) preserv'd the Radical Moisture in its due Vigor and Strength."[86] This view of the Law as a spiritual food that sustains the physical and moral health of the righteous can be fruitfully paralleled with the understanding of Hooker, Mather, and Herbert, among others, that gospel is the spiritual milk which the Protestant mother-minister, the "breast of God," feeds to his hungry flock of penitent sinners. Drawing on both the new science and on such Hebraic models of nursing motherhood as Sarah and Jochobed (the mother of Moses), the guidebooks promulgate a similar perception of the power of breast-milk to shape the moral and physical character of children in accordance with natural law and God's will. By turning to Hebraic exemplars and stressing likenesses among Sarah,

---

[84] John Lyly, *Euphues* (1580), *The Complete Works of John Lely,* ed. R. Warwick Bond (Oxford: Clarendon Press, 1902), 2.205.

[85] Rosenblatt, *Torah and Law in Paradise Lost,* pp. 25–6.

[86] Menasseh ben Israel, *De Termino Vitae libri Tres* (Amsterdam, 1639); *Of the Term of Life,* trans. Thomas Pocock (London: J. Nutt, 1699), p. 45. Cited in Rosenblatt, *Torah and Law in Paradise Lost,* p. 26.

Jochobed, and Mary, the guidebooks emphasize continuities between dispensations, between the life-sustaining properties of the Law and the regenerative power of mothers' milk, Mary's milk, and the milk of the Word.

Unlike the guidebook authors, however, Milton, in *Areopagitica*, remains ambivalent about whether or not law and gospel, nature and grace, Moses and Christ (who, as noted in Part 1 of this chapter, are both male nurses) complement one another as maternal fonts of nourishment. This ambivalence finds especially clear expression in Milton's comparison of books to food, in which he evokes three Pauline verses (Acts 10:13, 1 Thes. 5:21, and Titus 1:15) that abolish Jewish dietary law. Milton builds up to his extended Pauline book-food analogy by first commenting on St. Jerome's account of his "lenten dream" (*CPW*, 2: 510) that God had whipped him because he professed to love both Cicero and Christ. St. Jerome's vision was a matter of some controversy for those, like Paolo Sarpi, who debated whether or not the saint's whipping was a diabolical or angelic act. Detecting in Jerome's dream the "same political drift" that had caused "*Julian* the Apostat" ("suttlest enemy to our faith") to make "a degree forbidding Christians the study of heathen learning," Milton comes down in this debate on the side of those favoring "the Divell"— although he also more skeptically suggests that Jerome's dream was but "a fantasm bred by the fever which had then seis'd him" (*CPW*, 2: 508, 510). Had "an Angel bin his discipliner," Milton writes, he not only would have allowed St. Jerome to read Cicero (while justifiably chastising him "for dwelling too much upon Ciceronianisms"), but he would have also "let so many more ancient Fathers wax old in those pleasant and florid studies without the lash of such a tutoring apparition." Milton's contrast between the Angel's educational tolerance and the ghostly schoolmaster's devilish "lash" (2: 510) hearkens back to his concerns in *Of Education* with the differences between "happy nurture" (2: 377), which advances learning, and the distasteful education diet that impairs the intellectual health of students.

But this same contrast between gentle and harsh schoolmasters also paves the way for the turn that Milton makes in *Areopagitica* towards Pauline doctrine. Setting Jerome's dream against Eusebius's account of Dionysius Alexandrinus's "far ancienter" dream vision, Milton narrates how Dionysius "fell into a new debate with himselfe" after having been chastised by a "certain Presbyter" for his practice of attacking heretics by first "being conversant in their Books." Dionysius's conflict is resolved "when suddenly a vision sent from God … confirm'd him in these words: Read any books what ever come to thy hands, for thou art sufficient both to judge aright, and to examine each matter." Dionysius readily assents to this revelation because it echoes the verse of "the Apostle to the Thessalonians, Prove all things, hold fast that which is good" (*CPW*, 2: 511–12). Milton adds two Pauline verses to the one that Dionysius cites—both of which abolish the Jewish dietary prohibitions. Alluding to Titus 1:15, Milton writes:

> And he might have added another remarkable saying of the same Author; to the pure all things are pure, not only meats and drinks, but all kinde of knowledge whether of good or evill; the knowledge cannot defile, nor consequently the books, if the will and conscience be not defil'd.

Milton then cites Acts 10:13 to demonstrate that in the new dispensation, "mans discretion" rather than the Law determines his intellectual diet: "For books are as meats and viands are; some of good, some of evill substance; and yet God in the unapocryphall vision, said without exception, Rise *Peter*, kill and eat, leaving the choice to every mans discretion" (*CPW*, 2: 512). Milton's specific point is that Paul's overturning of laws that constrain men's choice of meat and drink authorizes Parliament to call in its Licensing Order, which similarly restricts men's choice of intellectual foods. But the broader impact of his Pauline allusions is that they characterize Hebraic law as adverse to life and liberty—as decidedly not-nurturing. Milton places the Law of Moses alongside the whipping schoolmasters, the inquisitorial priests and presbyters, and the fawning Florentine chancellors that he associates elsewhere in the tract with the thwarting of liberty of conscience and inward authority.

But if these parallels between the Law and other outmoded forms of punitive patriarchal authority help Milton to make his case against the parliamentary Licensing Order, they also detract from the close identification of England with the Israel of Hebrew scripture that he makes elsewhere in the tract: "Why else was this Nation chos'n before any other, that out of her as out of *Sion*, should be proclam'd and sounded forth the first tidings and trumpet of Reformation to all *Europ*" (2: 552). Milton's negative characterization of the Law also works against his efforts to model his own speech-act in *Areopagitica* on that of the Hebrew prophets, especially "*Moses* the great Prophet," the nursing father who cajoled and exhorted his reluctant people to embrace liberty rather than return to Egyptian bondage. Thus, even as he turns to Paul to shore up his argument, Milton also partly takes back his Pauline denunciation of the Law.

Milton makes this retraction when he comments on the tale in Exodus 16 of how God "tabled the Jews from heaven" (2: 513). Rather than oppose the Law to life and liberty, Milton demonstrates that the Law, like manna, nourishes both the body and the soul. Through the "Omer" that God (as nursing father) commanded Moses to supply as a daily ration to his people, the Hebrews were taught to temper their unruly appetites, their extravagant hunger for both "unwholesome" meat and false gods. The small "daily portion of Manna," Milton writes, is "computed to have bin more then might have well suffic'd the heartiest feeder thrice as many meals." "How great a vertue is temperance," he exclaims, "how much of moment through the whole life of man!" (*CPW*, 2: 513) Through a properly computed diet, Moses transforms the once-enslaved Hebrew people into a self-disciplined and self-governing nation. Whereas Milton opposes the Law to nurture and reformed nationhood when he alludes to Paul's epistles, he reconciles Hebraic law, Christian *caritas*, and national renewal in his evocation of Exodus 16 as a tale that reflects and sanctions his vision of the relation between nurture, mother-love, and Reformation. As *Areopagitica* demonstrates, Milton's highly ambivalent relationship to the Hebraic profoundly shapes his rendering of the relations among nurture and the nation. This is especially true of *Paradise Lost* and *Samson Agonistes*, as we shall see in the next two chapters.

Despite their manifest differences, then, all the examples examined in this chapter demonstrate the ways in which the language of maternal nurture seeps into and strengthens the self-fashioning of male political authority. By different means, James I, Charles I, Cromwell, and Milton first fill the reader/subject's heart with mother-love and nurture and then make political capital from these aroused filial affections and attachments.

4

Old Fathers and New Mothers:
Supersession and the "unity of spirit"
in *Paradise Lost*

Toward the end of Book 12, Adam descends from his hilltop scene of instruction with Michael to be reunited with the sleeping Eve. Not privy to the Archangel's lessons about reading human history typologically, Eve nevertheless has gained prophetic insight into the salvific future through the "gentle dreams.../Portending good" (12: 595–6) that Michael has sent her. Michael assures Adam that Eve's dreams have "calmed" her and "all her spirits composed/To meek submission" (12: 596–7). Her sleep has been creative: "As once thou slept'st, while she to life was formed" (11: 369). At Michael's command, Adam goes to awaken Eve, "but found her waked" (12: 598). He is left in the awkward position of being unable to execute the Archangel's commission—possibly a sign of Adam's waning authority (about which I shall have more soon). Eve welcomes Adam, but she also tells him that she already knows all about his comings and goings: "Whence thou return'st and whither went'st I know" (12: 610). She also informs her husband that she will play a pivotal role in humankind's redemption: "By me the promised seed shall all restore" (12: 623). Although she happily submits that Adam is to her "all things under Heav'n" (12: 618), she also commands him to "now lead on" (12: 614): "In me, there is no delay" (12: 615). Both Adam and Eve are described as "ling'ring" (12: 638) at Eden's eastern gate, but she appears more eager than he to begin their new life after their expulsion from Paradise.

The equivocal syntax and word order of Eve's speech also reflect the new ambiguities in her postlapsarian relationship to Adam. She acknowledges that, for her, Adam is the whole world "under Heav'n" (12: 618). She nevertheless cherishes the special superlunary "favor I unworthy am vouchsafed,/By me the promised seed shall all restore" (12: 622–3). Adam may be "all things" (12: 618) and "all places" (12: 618) to Eve, but he has neither place nor purpose in her last line. The ultimate "all" (12: 623) in Eve's speech and in Christian history will be restored "By me" (12: 623).

Although the first part of Eve's speech emphasizes her wifely subservience to Adam, the last four lines focus on her salvific identity as humankind's maternal redeemer, underscored by the adjacency between "me" and "the promised seed" in her last line (12: 623).[1]

Before ascending the hill in Book 11, Michael tells Adam that he has "drenched [Eve's] eyes" and put her to sleep "Here...below" (11: 366–8). Although we learn nothing about the specific contents of Eve's dream, her last speech in Book 12 offers some intriguing clues. Eve knows everything that Adam does before he returns to awaken her. This situation sharply contrasts with Book 8, where she chooses to withdraw from the conversation between Adam and Raphael to tend to her nursery of flowers, because "Her husband the relater she preferred/Before the angel" (8: 52–3). On this occasion, however, Eve prefers her angelic or divine relater before her husband, who remains silent after she narrates her dream. Adam either chooses not to speak, or he is prevented from speaking. In different modes, Adam and Eve learn the same lesson—he, by Michael's formal instruction, and she, more gently, through her dreams: "For God is also in sleep and dreams advise" (12: 611). As Michael teaches Adam, this lesson—*the* central lesson of Books 11 and 12—concerns the "better cov'nant" (12: 302) into which humankind will enter in "full time" (12: 301),[2] or when

> disciplined
> From shadowy types to truth, from flesh to spirit,
> From imposition of strict laws to free
> Acceptance of large grace, from servile fear
> To filial, works of law to works of faith.
> (12: 303–7)

One strong emphasis in this passage is on mediation and translation: on "from" (stated three times) to "to" (also stated three times). Although we don't know exactly what Eve sees or foresees in her dreams, we can safely surmise that her vision teaches her about her typological role as "second Eve" (5: 387) or Mary, Mother of Jesus, especially as a nurturing maternal mediator.

Eve happily masters her dream-lessons, at least partly, as we shall see, because typology enhances her importance as a mother in spiritual terms. Eve foresees that her typological translation into second Eve will reverse the seemingly irrevocable consequences of the Fall, attributed to her

---

[1] Louis Schwartz argues that Milton "recognizes the centrality of childbirth in working out the consoling plan of providence." *Milton and Maternal Mortality* (Cambridge: Cambridge UP, 2009), p. 234.

[2] My debt here is to John Rogers, *The Matter of Revolution: Science, Poetry, and Politics in the Age of Milton* (Ithaca: Cornell University Press, 1996), p. 177. Rogers conjectures that Eve's dream might offer a liberal counterpart to Michael's authoritarian pedagogy in Book 12.

terrible original lapse. Her expansive, charitable vision of supersession additionally prompts the hopeful authority that she expresses in her last speech. By contrast, Adam struggles with Michael's lessons on Christian typology, in part, because under this interpretive system his historical-and-material value as a father depreciates significantly. As presented in *Paradise Lost*, typology appears to reverse the traditional relations between gender and value: biological fathers decline, while spiritual mothers ascend.

Let us turn back briefly to *The power of women*, discussed in Chapter 2, because the image helps to illuminate the surprising ascendency of the spiritual/maternal and over the corporeal/paternal in Milton's epic. Two aspects are especially relevant to my discussion of Eve's female-centered typology. First, the new Protestant nursing mother replaces the old Holy Mother as mediatrix between the Hebraic dispensation and the reformed Christian era. Second, by nursing her babe at her breast, the new mother tenderly nurtures the supersession of the Old Testament by the New Testament (and of Roman Catholicism by Protestantism). The towering figure of the nursing mother occupies the center, consigning the two paradigmatic Hebrew men, Samson and Solomon (strength and wisdom), to the background. She allegorically associates the triumph of Christianity over Judaism with God's feeding hand (God as "Nourisher" in *Paradise Lost*). Supersession is life-enhancing, remedial, and maternal in spirit. Milton identifies supersession with spiritual maternal nurture through the shadowy figure of "second Eve" (5: 387), who is not yet, but nevertheless will be Mary, Mother of Jesus. We know this because Raphael supplies this identification in Book 5, when he "Hails" Eve as "blest Mary, second Eve" (5: 387).

Like Adam, Eve understands that a new covenant based on "free/ Acceptance of large grace" will replace the old one, grounded in the "imposition of strict laws" (12: 305, 304). Spirit rather than flesh will forge community. As we know from Book 9, Adam holds fast to "the bond of nature" (9: 956); therefore, for him, discipline will be central to his reformation from "flesh to spirit" (12: 304). By contrast, "works of faith" (12: 306) appear to be the happier emphasis in Eve's gentle vision of the epic's central prophecy. She foresees that, in her future role as divinely inspired maternal mediator, she will tenderly nourish the spirit of unity shared by all true citizens of the godly nation. Pauline doctrine reinforces Milton's typological distinction between male/flesh and female/ spirit, and between the embodied Israel of the outmoded past and the disembodied one of the regenerate future. Eve's "promised seed" (12: 623) alludes to Paul's distinction between "the children of the flesh" and "the children of the promise" from Romans 9:6–8:

For not all who are descended from Israel belong to Israel, and not all are children of Abraham because they are his descendants... This means that it is not the children of the flesh who are the children of God, but the children of the promise are reckoned as descendants.

In *De Doctrina*, Milton echoes Romans 9: "The promise, therefore, is not to the children of Abraham in the physical sense, but to the children of his faith who received Christ" (*CPW*, 6: 196). In *Paradise Lost*, Adam loses the paradisal promise of natural patriarchal preeminence after the Fall. Eden might have been Adam's "capital seat, from whence had spread/All generations, and had hither come/From all the ends of th'Earth, to celebrate/And reverence thee their great progenitor" (11: 343–6). The Fall, however, forever impairs this biologically fruitful, patriarchal future.

Whereas Adam's natural patriarchal preeminence is diminished by the Fall, Eve's spiritual powers of maternal nurture are enhanced by her typological translation into second Eve. We look forward to the Mary of *Paradise Regained* who nurses Jesus's typological insights into his divinely inspired role as humankind's savior. So close is the Mary–Jesus, mother–son dyad that each is attuned, without knowing it, to the other's thoughts. When Jesus fails to return after his baptism, Mary's "breast, though calm, her breast though pure/Motherly cares and fears got head and raised/Some trouble thoughts" (*PR*, 2: 63–5). Jesus's mental serenity is similarly disturbed in the desert: "O what a multitude of thoughts at once/Awakened in me swarm, while I consider/What from within I feel myself and hear/What from without comes often to my ears,/Ill sorting with my present state compared" (*PR*, 1: 196–200). Although physically separated from one another, mother and son regain spiritual composure simultaneously. She "with thoughts/Meekly composed awaited the fulfilling/The while her son tracing the desert wild,/Sole but with holiest meditations fed,/Into himself descended" (2: 107–11). Milton gracefully depicts the mind-melding intimacy between Mary and Jesus. Her "thoughts/Meekly composed" (2: 107–8) harmonize perfectly with the "holiest meditations" (2: 110) upon which her Son so satisfyingly feeds. Milton's Mary and Jesus model the spirit of unity and fortifying charity that will nourish the new Israel, when freed from the Roman yoke. Bonds of spirit will replace the bond of nature, which is the very same bond (and bondage) that Adam cannot break after Eve's transgression. The new spiritually unified Israel is glimpsed when Jesus returns "Home to his mother's house private" (4: 639), from which the biological father is excised and the materiality of flesh (maternal, paternal, and filial) is left behind.

If she resembles the nursing mother in *The power of women*, Milton's new mother significantly differs from the guidebooks' ideal. Rather than naturally transmit pure English identity and a reformed national ethos through her pure breast-milk, Milton's new mother affectively nourishes the unity of spirit that consolidates the reformed post-dynastic nation: "neighboring differences... need not interrupt *the unity of Spirit*" (*CPW*, 2: 565). In *A Needful Corrective and Balance in Popular Government*, Henry Vane, to whom Milton dedicated a sonnet in 1652, speaks analogously of the new spiritual citizen:

> None be admitted to the right and privilege of a free Citizen, for a Season, but either such as are free born, in respect of their holy and righteous principles, flowing from the birth of spirit of God in them (restoring man in measure and degree as at the first Creation) unto the right of Rule and Division.[3]

For Milton like Vane, reformed citizens are "free born" in spirit, happily released from the outmoded bonds of nature, blood, or kinship, and united in disembodied terms by shared righteous principles. Like Milton, Vane identifies free citizenry with rebirth in Christ and, so, also with the emancipating translation of the Old Testament into the New Testament—a threshold crossing that Milton associates with the nurturing, second Eve. Notably, for Vane, to be a free citizen is to be free of original sin: "restoring man in measure and degree as at the first Creation." With a female accent, the "all restore" (12: 623) of Eve's last line resonates with Vane's conception of restoration as the revival of a free, spiritually redeemed citizenry. Eve's emancipating restoration of "all" (12: 623) opposes the repressive Restoration of Stuart monarchy. The female alternity of her speech further underscores the politically subversive implications of her reference to "restor[ation]" (12: 623).

Eve's typological, maternal vision of human history as ultimately restorative sharply contrasts with what David Loewenstein aptly describes as the tragic "drama of history" that Adam witnesses in Books 11 and 12.[4] This tragedy includes "the Hebrew tragedy of Torah degraded into law and of redemption purchased at a terrible price," as Rosenblatt argues.[5] Whereas the epic's Christian comedy culminates with Eve's vision of the divinely inspired restoration of "all...lost" (12: 621), its Hebrew tragedy turns on the fallen Adam's acknowledgment of his obsolescence as a natural father and the fatal curse that the children of his loins and the biblical

---

[3] Henry Vane, *A Needful Corrective or Balance in Popular Government* (London, 1660), pp. 7–8. Vane's tract is addressed to James Harrington.

[4] See David Loewenstein, *Milton and the Drama of History: Historical Vision, Iconoclasm, and the Literary Imagination* (Cambridge: Cambridge University Press, 1990).

[5] Jason P. Rosenblatt, *Torah and Law in Paradise Lost*, p. 234.

Israel inherit from him. Adam shows us how lack, loss, despair, and weakness are constituted in the fallen nation and national subject.[6] He also does the work of mourning in the poem: mourning for the passing of the past, the perversion of Mother Nature, the demise of biological fatherhood, the cursed fate of natural progeny, the mortality of the body and body politic, and the supersession of the biblical Israel, in the Pauline terms favored by the last two books.

Through mourning, Adam also paradoxically anticipates Eve's joyous final speech. In accepting the Son as his savior, he models in masculine terms the feminine ways in which loss can be converted to gain, despair to joy, weakness to strength, the old father to the new nurturing mother and son. Although Adam fails to break the "bond of nature" (9: 957) or his visceral connection to Eve and his mother Earth, from whose clay his body is formed, Adam, under Michael's tutelage, ultimately recognizes that the "paradise within" is "happier far" (12: 587). By affirming the primacy of spirit and interiority—that is, nurture over nature, love over law—Adam also accepts that his autochthonous birth and special status as the world's first natural father are drained of value in the new era. Milton's depreciation of Adam's natural prestige echoes 1 Cor. 15:48: "The first man *is* of the earth, earthly; the second man *is* the Lord from heaven." In his allusion to the "second Adam" (11: 383), Milton may also hint at Origen, who denies the first Adam's historical reality altogether by translating the natural father into the spiritual Son, and the first Adam to the second Adam.[7] As Rosenblatt observes, "The old Adam must be fully absorbed in the second Adam, just as the Church requires the absorption of Judaism in Christianity, the second Israel."[8] Thus, for Milton, while Adam ultimately masters Michael's lessons on the regenerate future, he and his "race unblest" (10: 988) ultimately will be excluded from the "children of the promise," or of Abraham's faith, to whom the new Israel is bequeathed at the End of Days. As Michael teaches Adam in Book 12, the Savior brings salvation "Not by destroying Satan, but his works/In thee and in thy seed" (12: 394–5).

Descent is the crucial issue for Paul. By turning Abraham into the father of both the children of the flesh and the children of the promise, the apostle found an effective way to manage the fraught relations between the first Israel and the second one. Rather than Abraham's biological

[6] I discuss the melancholic Adam in "Sublime Pauline: Denying Death in *Paradise Lost*," in *Imagining Death in Spenser and Milton*, ed. Elizabeth Jane Bellamy, Patrick Cheney, and Michael Schoenfeldt (New York: Palgrave, 2003) pp.131–50.

[7] On Origen, I refer to David Somerville, M.A., *St. Paul's Conception of Christ or The Doctrine of the Second Adam* (Edinburgh: T&T Clark, 1897), p. 52.

[8] Jason P. Rosenblatt, *Torah and Law in Paradise Lost*, p. 58.

offspring, the true descendants of Abraham are the children of the faith who embrace the promise:

> For not all who are descended from Israel belong to Israel, and not all are children of Abraham because they are his descendants... This means that it is not the children of the flesh who are the children of God, but the children of the promise are reckoned as descendants (Romans 9: 6–8).

"Know ye therefore that they which are of faith, the same are the children of Abraham" (Gal. 3:7). Milton's Abraham resembles and expands upon Paul's Abraham. In *Paradise Lost*, Michael teaches Adam that, through Abraham, the ascendancy of second, spiritual children over first, natural offspring finds sanction. Abraham also offers scriptural justification not only for the primacy of the new England/Israel over the old corporealist or dynastic nation, but also for the supersession of abstract, spiritual, and affective ties over blood and genealogical bonds. Eve receives a similar, but gentler vision of the new unity of spirit in her dream, in which the reformed mother, not the customary patriarch, plays a central, mediating role.

Rather than to Adam and his "hapless seed" (10: 965), the way to the salvific new Israel is revealed to Abraham's sons, both his biological and his spiritual progeny: "Not only to the sons of Abraham's loins/Salvation shall be preached, but to the sons/Of Abraham's faith, wherever through the world" (12: 447–9). As Rosenblatt observes, the "not only...but to" in these lines unites the "sons of Abraham's loins" with the "sons of Abraham's faith" as members of the same elite but expansive community of those to whom "Salvation shall be preached."[9] But, although Milton allows both Abraham's biological sons and Abraham's spiritual sons to hear the good news, the poet does not specify whether or not "the sons of Abraham's loins" (12: 447) will answer the call. Elsewhere in his writings, the poet emphatically states that they will not. In *De Doctrina*, Milton specifies that only Abraham's spiritual sons, "the children of his faith who received Christ," will receive the promise and find redemption: "The promise, therefore, is not to the children of Abraham in a physical sense, but to the children of his faith who received Christ" (*CPW*, 6: 196). In *Paradise Regained*, Jesus denounces "those captive tribes" (*PR*, 4: 414) as an idolatrous race, "Unhumbled, unrepentant, unreformed" (*PR*, 329). The Son of God describes the call of the Jews in the conditional: "Yet he at length, time to himself best known,/Rememb'ring Abraham, by such wond'rous call/May [or may not] bring them back repentant and sincere" (*PR*, 3: 433–5).

---

[9] See *Torah and Law in Paradise Lost*, pp. 227–9.

In *Paradise Lost*, however, the relations between the Hebraic and the Christian are more ambiguous.[10] Hebrew tragedy and Christian comedy closely mesh in Milton's epic, as do Adam and Eve, and hence both are considered in detail in this chapter. The reformation of motherhood, or the typological translation of Eve to "second Eve" (5: 387, 10: 183), intertwines with the Pauline vision of the new spiritual Israel in Books 11 and 12. Reformation, however, cannot be achieved without the tragic experience of loss, sacrifice, and suffering, without paying a terrible price. For Maureen Quilligan, the terrible price of regeneration is Eve's submission and subjection to Adam: "to thy husband's will/Thine shall submit, he over thee shall rule" (10: 195–6).[11] As we have seen, Adam also pays a terrible price. In the end, Milton partly compensates Eve for her wifely subjection to Adam by translating her as "second Eve" (5: 387, 10: 183) into the maternal agent of humankind's deliverance—a role appointed to her alone.[12] By contrast, Adam ultimately is a diminished figure. Once "no shadow" (4: 470), he devolves into a "shadowy typ[e]" (12: 303), an emptied-out, silent emblem of the attenuated body of the Hebraic past and the Israel of Hebrew scripture, which is reduced to the status of "captive text."[13] Even so, in the end, Milton is unable to consign Adam and Israel to full oblivion. Although Adam ultimately sacrifices the primacy of the body for that of the spirit, his strong attachment to his embodied self remains palpable. Michael teaches Adam to exchange the Hebraic monism of the lost Paradise for the Pauline dualism that governs postlapsarian thought and experience. At the same time, however, Milton's materialist monism, Vitalist view of nature, and mortalism[14]—alluded to in the poet's reference to soul-sleeping (the "death-like sleep" [12: 435]) of the

---

[10] Achsah Guibbory brilliantly discusses the broader cultural context of Christian–Jewish relations in *Christian Identity, Jews, and Israel in Seventeenth-Century England* (Oxford: Oxford University Press, 2010).

[11] Maureen Quilligan, *Milton's Spenser: The Politics of Reading* (Ithaca: Cornell University Press, 1983), p. 237. In "The Genesis of Gendered Subjectivity in the Divorce Tracts and in *Paradise Lost*," in *Re-membering Milton: Essays of the Texts and Traditions*, ed. Mary Nyquist and Margaret Ferguson (New York: Methuen, 1987), pp. 99–127, Mary Nyquist argues that Eve is educated in submissiveness. See also James Grantham Turner, *One Flesh Paradisal Marriage and Sexual Relations in the Age of Milton* (Oxford: Clarendon, 1987).

[12] In *The Reformation of the Subject: Spenser, Milton, and the English Protestant Epic* (Cambridge: Cambridge University Press, 1995), Linda Gregerson argues that Eve's postlapsarian subjection makes her "the normative, postlapsarian human subject," who illuminates the "paths available to humankind" (pp. 196–7). In *Ceremony and Community*, Guibbory argues that "Woman is implicitly charged with the primary responsibility not only for the Fall but also for effecting the right relation of human and God," p. 218.

[13] Rosenblatt, Torah and Law in *Paradise Lost* (Princeton: Princeton University Press, 1993), p. 63.

[14] On Milton's Vitalism and materialist monism, see Rogers, *The Matter of Revolution: Science, Poetry, and Politics in the Age of Milton* (Ithaca: Cornell University Press, 1996), Chapters 4–5.

regenerate—also re-identifies the spiritual with the material in reformed Christian terms.

Paradise Lost's tragicomic, Judeo-Christian, and paternal–maternal vision of the reformed Israel-England clearly necessitates a multifaceted critical approach. I argue that an assessment of the role that second Eve plays as reformed mother in the spiritual renewal of Israel is best made through a feminist analysis of the status of the maternal and of female authority and subjectivity in the specific historical and cultural context of the 1640–60s. I also argue that Adam's tragic role as the outmoded father of the old natural nation and its "race unblest" (10: 989) cannot be appreciated fully without taking into account many of the insights that Jewish-studies scholars have made into the sexual and cultural politics of Judeo-Christian relations.[15] In the pages that follow, I make an inaugural attempt to read Milton's epic from the very different but, I believe, mutually informing vantage points of early modern feminist literary studies and Jewish studies. In this way I hope to shed new light on the important but hitherto obscure affinities in Milton's epic among the new mother, supersession, and the new Israel. In its most ambitious scope, this chapter attempts to provide a new bridge between the gender and religio-national concerns of *Paradise Lost*.[16]

EVE ON TOP

While sometimes relegated to an insignificant postscript to Adam's extended tutorial with Michael in Books 11 and 12, Eve's speech, as I have just suggested, provides an important point of departure for analyzing the interlocking relations between mother- and nationhood in Milton's epic and the central role that reformed motherhood plays in the poet's vision of national redemption. Milton's strategic placement of the speech at the very end of his epic accentuates the political implications of its gendered inversions, forward-looking utopianism, iconoclasm, and maternal-centered vantage point on providential history and the revealed truth. With Eve, we leave Eden with the subversive image of the world-turned-upside-down and a spiritual mother on top. For Milton, the refilling of

---

[15] See Daniel Boyarin, *Carnal Israel*. As applied to *Paradise Lost*, see *Torah and Law in Paradise Lost* and Jeffrey Shoulson, *Milton and the Rabbis* (New York: Columbia University Press, 2001).

[16] Guibbory argues that interlocking tensions between seventeenth-century English perceptions of female and Jewish identity shape Milton's thoughts about the nation in *Samson Agonistes* in "The Jewish Question" and "The Woman Question in *Samson Agonistes*: Gender, Religion, and Nation," in *Milton and Gender*, ed. Catherine Gimelli Martin (Cambridge: Cambridge UP, 2004), pp. 184–206.

the world with hope, gratitude, exaltation, and prophetic expectancy comes *after* the tragic drama of history that Adam witnesses in the epic's final two books, and it has everything to do with both the spiritual renewal of motherhood and the end of the natural paternal line and dynastic progression. At the end of the epic, Adam, our "great progenitor" (11: 346) neither has lines to speak (Adam "answered not" [12: 625]) after Eve's speech nor a healthy line of biological descendants. As Michael observes, had Adam not fallen, he would have been revered as the father of all future generations: "All generations...had hither come/...to celebrate/ And reverence thee their great progenitor/But this preeminence thou hast lost" (11: 344–7). Against kingship, corporealism, defeat, and death, Milton links the eventual reformation of Eve's motherhood to the promised reformation of the nation at the End of Days, after the spiritual rebirth of nature and the body.

One especially rich historical analogue to Eve's final visionary speech can be found in the prophetic writings of seventeenth-century sectarian women. Phyllis Mack observes that between 1640 and 1660, there were "over 300 women visionaries of whom about 220 were Quakers"; she documents that "Over four hundred women prophesied at least once during the second half of the seventeenth century."[17] Women's prophetic writing equates social renovation, revelation, and the new mother in much the same way that Milton associates second Eve with the revealed Truth and the spiritually unified nation. As I have argued elsewhere, Eve's inversionary dream vision and future-directed rhetoric recalls the female prophetic writing that proliferated in the middle decades of the seventeenth century and the new public visibility and political authority that women achieved as prophetic authors and speakers.[18] Katherine Romack observes that female prophecy "has a distinctly feminized poetics, complete with its own formal conventions, intimately connected to women's social and political place in early modern culture, and crucially characterized by...gendered inversions."[19] Eve's speech does not replicate all of the formal conventions of female prophecy, specifically, its purposefully

---

[17] Phyllis Mack, "Women as Prophets During the English Civil War," *Feminist Studies*, 8.1 (1982): 24; "The Prophet and Her Audience: Gender and Knowledge in the World Turned Upside Down," in *Reviving the English Revolution: Reflections and Elaborations on the Work of Christopher Hill*, ed. Geoff Eley and William Hunt (London: Verso, 1988), p. 150.

[18] Rachel Trubowitz, "Female Preachers and Male Wives: Gender and Authority in Civil War England," in *Pamphlet Wars: Prose in the English Revolution*, ed. James Holstun (London: Frank Cass, 1992), pp. 112–33.

[19] Katharine Romack, "Monstrous Births and the Body Politic: Women's Political Writings and the Strange and Wonderful Travails of Mistris Parliament and Mrs. Rump," in *Debating Gender in Early Modern England, 1500–1700*, p. 212.

incoherent, stuttering style. Nevertheless, her speech is marked by the same "multivalenced revelatory logic," gendered inversions, and woman-on-top rhetoric characteristic of seventeenth-century women's visionary writings.[20] It is "By me" (12: 623), Eve declares, that the Fall will be reversed. By the 1660s, sectarian women prophets had faded into oblivion along with the revolutionary prospect of a new anti-monarchical order, of which they were an emblem. By turning Eve into a prophet and by underscoring her central role in humankind's redemption, however, Milton remembers the alternative or "female" national future heralded by the Good Old Cause—a vision struck from the official historical record by the Restoration's Act of Oblivion.

The maternal metaphors in Eve's regenerative vision of the future most closely identify her final speech with female prophetic rhetoric. Female reproduction and nurture are recurrent tropes in seventeenth-century prophetic writing.[21] Nursing serves as an important political metaphor. In *A Warning to the Dragon* (1625), Lady Eleanor Davies proclaims that God "powrth out his Spirit upon his hand-maiden...It is milke for the young." The city of "BabyLondon" or the daughter of London or Babylon/Rome (which also is, for Davies, the London of Charles I and Henrietta Maria) refuses natural maternal nurture and instead thirsts for monstrous "hart-Blood": "In steed of her Mothers milke and breast she sucks her hart-Blood, whose Father was a Lyer and a Murtherer from the beginning." With vivid, intimate detail, she proclaims that the Lord will "forget [the godly] no more then a Mother forgets her new-borne Sonne, her sucking childe, when the Tongue cleaves to the roofe of the Mouth for Thirst."[22] The Fifth Monarchist, Mary Cary, in *The Little Horns Doom and Downfall, and A New and More Exact Map* (1651), also describes God as a maternal nourisher: "And it shall come to pass in that day that...the Hils

---

[20] Important studies of women prophets during the revolutionary decades include: Keith Thomas, "Women and the Civil War Sects," *Past and Present*, 13 (1958): 42–62; Nigel Smith, *Perfection Proclaimed: Language and Literature in English Radical Religion 1640–1660* (Oxford: Clarendon Press, 1989), esp. 45–53; Phyllis Mack, *Visionary Women: Ecstatic Prophecy in Seventeenth-Century England* (Berkeley: University of California Press, 1992); Susan Wiseman, "Unsilent Instruments and the Devil's Cushion: Authority in Seventeenth-Century Women's Prophetic Discourse," in *New Feminist Discourse: Critical Essays on Theories and Texts*, ed. Isobel Armstrong (London: Routledge, 1992); Sharon Achinstein, "Women on Top in the Pamphlet Literature of the English Revolution," *Women's Studies* (1994): 131–64; and Shannon Miller, "Milton Among the Prophets: Inspiration and Gendered Discourse in the Mid-Seventeenth Century," in *Engendering the Fall: Milton and Seventeenth-Century Women Writers* (Philadelphia: University of Pennsylvania Press, 2008), pp. 79–106.

[21] As Stevie Davies observes, "all of this panging was perfectly in keeping with a prophetess's necessary birth-throes," *Unbridled Spirits: Women of the English Revolution, 1640–1660* (London: The Women's Press, 1998), p. 141.

[22] Lady Eleanor Davies, *A Warning to the Dragon* (London, 1625), A3v–A4r.

shall flow with milk." Weaving together passages from Isaiah, Daniel, and Revelation, she writes:

> And thus now it is cleer in these several Scriptures...that the Gentiles shall be plentifully and abundantly brought into the obedience of Christ...and not onely the mean and inferiour sort of the people of the Nations, but some of thir Kings and Queens...Kings shall be their nursing fathers and Queens their nursing mothers.

Cary also relies on metaphors of mothers' milk, breasts, and nursing to demonstrate that Christ will confer both outward and inward glory on the New Jerusalem:

> These are the things wherein the external glory of a land consists; and he says withal, that they shall suck the milk of the Gentiles, and such the breasts of Kings. Which imports that the treasures, and outward glory of the Nations, and of the Kings, and great men of the world, shall be all drain'd together, and brought in to them they shall suck the sweetness of them, and shall injoy them all. And thus will the Lord make his Sion an external excellence in outward respects, as well as in inward or spiritual respects.

Cary dedicates her tract to Lady Elizabeth Cromwell, Lady Bridget Ireton (Cromwell's daughter), and Lady Margaret Role because she feels that these women will understand and embrace her millennial vision:

> I have therefore chosen, (because of your own sex) to dedicate these Treatises to your Ladyships being assured both, First of you ingenuous and gracious acceptation hereof...And also secondly, of your owning, defending, and maintaining all the truths, which are therein laid down.[23]

These dedications, as Rachel Warburton argues, "construct a community of female readers for Cary's visions."[24] I would add that Cary hails this same readership through her vivid allusions to motherhood and maternal nursing.

---

[23] Mary Cary, *The little horns doom & downfall or A scripture-prophesie of King James, and King Charles, and of this present Parliament* (London, 1651; Thomason/E.1274[1]), sig. M4v, p. 168, sig. T8r, p. 287, sig. A3r.

[24] Rachel Warburton, "Mary Cary (Rande), *Little Horns Doom and Downfall and a New and More Exact Mappe of the New Jerusalems Glory*" (1651), in *Reading Early Modern Women: an Anthology of Texts in Manuscript and Print 1550–1700*, eds. Helen Ostovich and Elizabeth Sauer (New York and London: Routlege, 2004), p. 157. Jane Baston compares Mary Cary's vision to a lost "utopian socialism," in "History, Prophecy, and Interpretation: Mary Cary and Fifth Monarchism," *Prose Studies*, 21.3 (December 1998): 6. Katharine Gillespie reads Mary Cary as a female precursor to John Locke, "Improving God's Estate: Pastoral Servitude and the Free Market in the Writings of Mary Cary," *Domesticity and Dissent in the Seventeenth-Century: English Women Writers and the Public Sphere* (Cambridge: Cambridge UP, 2004), pp. 167–215.

In *The Cry of a Stone*, Anna Trapnel (like Cary, a Fifth Monarchist) looks forward to the time after the trials and tribulations of world history, when God's poor suffering saints will finally rest in eternal peace. Not unlike Milton's Eve, Trapnel emphasizes the maternal, nurturing aspects of this afterward and new beginning. For Trapnel, the ultimate place of rest is Jesus's nurturing breast, the nest wherein he nourishes his hungry people and shelters them from danger. "They lodge within that breast," she proclaims, "they rest and nest in Jesus Christ."[25] Trapnel expresses these sentiments in verse:

> What are the birds and cattel there
> Whereon man is fed?
> Thou [Christ] are a rest to weary man,
> Who forth and in is led...
> Of all things they [his Saints] shall draw the sap
> That runneth from the root,
> And get up into a high tree
> Where none shall pluck
> No, none shall be above to see
> Thine, when th'are in the nest
> For they are closed in so round
> They lodge within that breast.
>
> (sig. H1r, p. 49)

The simple, *gemütlich* rhymes among "rest," "nest," and "breast" reinforce Trapnel's linkage of Christian redemption with maternal and domestic comfort, sustenance, and love.

As these recurrent tropes of maternal nursing suggest, prophetic discourse constitutes the visionary speaking subject in female terms. As Phyllis Mack observes, many Protestants "depicted prophecy as a feminine activity, whether the actual prophet was a man or woman."[26] Abiezer

---

[25] Anna Trapnel, *The cry of a stone, Or a relation of something spoken in Whitehall* (London, 1654). Important studies of Trapnel include David Loewenstein, *Representing Revolution in Milton and His Contemporaries: Religion, Politics, and Polemics in Radical Protestantism* (Cambridge: Cambridge University Press, 2001), pp. 116–24; James Holstun, "The Public Spiritedness of Anna Trapnel," *Ehud's Dagger: Class Struggle in the English Revolution* (London: Verso, 2000), pp. 257–304; Hilary Hinds, *God's Englishwomen: Seventeenth-Century Radical Sectarian Writing and Feminist Criticism* (Manchester: University of Manchester Press, 1996), pp. 122–8; and Diane Purkiss, "Producing the Voice, Consuming the Body: Women Prophets of the Seventeenth Century," in *Women, Writing, History, 1640–1740* (Athens, Georgia: University of Georgia Press, 1992), pp. 139–58.

[26] Phyllis Mack, "Women as Prophets," p. 24. Elizabeth Sauer alternatively distinguishes male from female forms of prophecy in "Maternity, Prophecy, and the Cultivation of the Public Sphere in Seventeenth-Century England," *Renaissance Culture*, 24 (1998): 118–48. Shannon Miller argues that Milton draws on the distinct stylistic differences between male and female prophets in his characterizations of Adam and Eve, *Engendering the Fall*, pp. 86–91.

Coppe, the Ranter leader, compares his words to "the child in the womb I was so big of."[27] Like Coppe, Puritan ministers, as we have seen, cultivated qualities associated with wives and mothers in their posture towards the deity and their congregations. In order to become fully receptive to divine influence, they acquired a metaphorical female openness and interiority so that they could be filled by inner light. It bears repeating that Puritan ministers routinely depicted themselves as breasts of God and their congregations as newborn babes desiring the milk of the Word (Lev. 146). John Cotton maintains that ministers should "apply themselves to the estate of their people: If they bee babes in Christ, to be as breasts of Milke to suckle them."[28] Thomas Hooker glosses Isaiah 66.11 as saying,

> The Church is compared to a childe, and the brests are the promises of the Gospell; now the elect must suck out and be satisfied with it, and milke it out…Ah beloved, this is our misery, we suffer abundance of milke to be in the promise, and we are like wainly children that lye at the brest, and will neither sucke nor be quiet.[29]

Puritan sermons also highlight the new importance that the metaphorical description of the church as a wife in Ephesians 5:23—"For the husband is the head of the wife, even as Christ is the head of the church"—assumes in the conception and practice of Puritan devotion. As Maureen Quilligan points out:

> If each believer had become his own priest, and was no longer a member of an institutionally visible church, the priest found his congregation had shrunk to the literal foundation upon which Paul had based his metaphoric description of Christ's relationship to his church: the love of a husband for his wife.[30]

In spiritual terms, both men and women are wives of Christ, the Bridegroom. "Christ," writes Margaret Fell Fox, "is the Husband, to the Woman as well as the Man, all being comprehended to be the Church."[31] By adopting the wifely posture of submission to God, Puritans internally literalize the Pauline metaphor of the church as beloved helpmeet. Adam adopts this submissive, wifely role at the end of Book 12 when he proclaims his devotion to God as an obedient, physically weak but spiritual strong "female" disciple:[32]

---

[27] Abiezer Cope, *A Second Fiery Flying Roll* (London, 1650), p. 14.
[28] John Cotton, *A Brief Exposition of the whole Book of Canticles…* (London, 1642), pp. 240–1.
[29] Thomas Hooker, *The Soules Vocation* (London, 1638), p. 306.
[30] Maureen Quilligan, *Milton's Spenser*, p. 224.
[31] Margaret Fell Fox, *A Call to the Universal Seed of God* (London, 1665), pp. 16–17.
[32] See Albert C. Labriola, "The Aesthetics of Self-Diminution: Christian Iconography and *Paradise Lost*," *Milton Studies*, 7 (1975): 267–311.

> Henceforth I learn, that to obey is best
> And love with fear the only God, to walk
> As in his presence, ever to observe
> His providence, and on him sole depend,
> Merciful over all his works, with good
> Still overcoming evil, and by small
> Accomplishing great things, by things deemed weak
> Subverting worldly strong, and worldly wise
> By simple meek; that suffering for truth's sake
> Is fortitude to highest victory.
>
> (12: 562–70)

Adam ultimately will subvert "worldly strong, and worldly wise" by "suffering for truth's sake." To achieve "highest victory" through suffering, however, Adam must first be taught by Michael to reorient himself in submissive, female terms toward God. By contrast, Eve needs no lessons on gender reorientation from Michael. In her final visionary speech, she exhibits the inspired maternal vantage point required to "milke... out" the "brests [that] are the promises of the Gospell."[33]

Puritanism offered women and other socially deprived groups the power "to trump the authority of men based on the education they had monopolized," as John Strachniewski observes.[34] In opposition to official truths, prophetic discourse celebrates the unorthodox, alternative, and female-coded insights that were to bring salvation to the regenerate Protestant subject. Digger leader Gerrard Winstanley in *The Law of Freedom in a Platform* writes that children should "not be trained up only to book learning and no other employment, called scholars, as they are in the government of monarchy."[35] To undermine this surge of intellectual power from below, critics belittled Puritan emphasis on the free way rather than the formal way to acquire knowledge as lacking in masculine rigor. In *Enthusiasm Triumphatus*, Henry More argues that women's stubborn refusal to yield to formal logic and argumentation typifies puritans' "intellectual intransigence" and political opposition to the governing social hierarchy. Argue with a woman, writes More, and "she will look upon it as a piece of humane sophistry, and prefer her own infallibility or the infallibility of the Spirit before all carnall reasonings whatsoever."[36]

---

[33] Thomas Hooker, *The Soules Vacation* (London, 1638), p. 306.

[34] John Strachniewski, *The Persecutory Imagination: English Puritanism and the Literature of Religious Despair* (Oxford: Clarendon, 1991), p. 43, and esp. pp. 40–4.

[35] Gerrard Winstanley, *The Works of Gerrard Winstanley*, ed. George Sabine (Ithaca: Cornell University Press, 1941), p. 577.

[36] Henry More, *Enthusiasm Triumphatus: Or, a Discourse on the Nature, Causes, Kinds, and Cure of Enthusiasm* (London, 1656), p. 57. Cited in John Strachniewski, *The Persecutory Imagination: English Puritanism and the Literature of Religious Despair* (Oxford: Clarendon, 1991), p. 43.

By contrast, guidebook authors make positive, productive use of the connection between women and Puritanism's repudiation of lettered authorities in their exhortations to entrust the education of young children to mothers. In his *Briefe Explanation*, Cleaver urges mothers to instruct children in the milk of the Word while they are nursing. Children should become naturally fluent in the language of the gospels from the breast: St. Paul "would have them to sucke in religion, if not with their mothers milke, or shortly after as-soone as they are capable of it."[37] Milton's characterizations of Adam and Eve as learners exhibit a similar kind of gendered conflict between lettered and unlettered knowledge. Adam displays the intellectual limitations and conservatism of the formally trained exegete. Clinging to the old order, he errs time and again in his efforts at interpreting the biblical scenes that Michael shows to him: he adheres too closely to the letter—a point that I shall examine more closely toward the end of this chapter. Unlike Adam, Eve bypasses the letter entirely, trusting entirely in the infallibility of her divinely inspired dreams of the "great good/Presaging" (12: 612–13).

As already noted, Eve proclaims to Adam that "thou to mee/Art all things under Heav'n" (12: 617–18). She, however, also emphasizes at the end of her speech that her role as a providential agent of universal redemption is not as Adam's spouse, or, as first Eve, but rather as second Eve or Virgin Mother. Her radical separateness from her human spouse in the new era is anticipated by the ways that her final speech corrects her primal tendency toward self-absorption, which Eve betrays in her account of her nativity in Book 4. As in her earlier narrative, when she describes how she became enamored by her own image reflected in the "liquid plain" (4: 455), Eve in her last speech is focused on "me" (12: 623). In Book 4, she tells how she was guided by God's voice away from her own shadowy, liquid reflection to Adam, who is "no shadow" (4: 470), but rather both her original ("whose image thou art" [4: 472]) and her substantial "other half" (4: 488). By contrast, in her final speech, she is instructed by God through her dream to see herself perfected not by Adam, a shadowy type, but rather by her redeemed future self and typological fulfillment as "blest Mary, second Eve" (5: 387).

In political terms, the regenerative message of Eve's final speech is this: before the Fall, sexuality, marriage, and the production of natural progeny were to establish kinship, bloodline, and genealogy as divinely sanctioned measures of personal and communal identity; after the Fall, these natural, procreative responses to God's first law, to be fruitful and multiply, devolve

---

[37] Robert Cleaver and John Dod, *A Briefe Explanation of the Whole Booke of the Proverbs of Salomon* (London: Felix Kyngston, 1615), pp. 352–3.

into sin and death. After the Fall, only spiritual means of producing progeny, such as evangelism, baptism, and conversion, can generate the spiritual unity that is the cornerstone of a free citizenry and godly nation.
This point is made in the last four lines of Eve's speech, which contain no
direct reference to Adam, even though they form the final statement of
her conjugal conversation with her spouse and the very last piece of
spoken dialogue in *Paradise Lost*. After Eve finishes speaking, Adam is
silent, partly at Michael's gentle insistence. As already noted, although
Adam is "Well pleased" (12: 625) by Eve's divinely inspired prophecy, he
nevertheless "answered not" (12: 625). His silence responds to the last
word that Eve utters: "restore" (12: 623). This is the reversal of "all...lost"
(12: 621) by the promised seed that Eve will deliver and nurture. Remarkably, Adam is granted no seminal role in this miraculous, epoch-shifting
transformation. As Michael painstakingly teaches him, Adam's regenerative passage from the old dispensation into the new era requires that he
give up the primacy of his role as biological father, a sacrifice that, until
the very end of the epic, Adam is unable to make.

David Loewenstein insightfully underscores the "austere," "unadorned,"
and "dark" dimensions of the "drama of history" presented in Books 11
and 12. Even Michael's repeated assurances of typological promise and
fulfillment cannot eradicate "the profound sense of history as an ongoing
process of tragic conflict...a tragic vision that so unrelentingly dramatizes the conflicts and terror of human history."[38] Eve's last speech, however, forms a surprisingly overlooked comic postscript to the tragic vision
of human history set out in the last books. Like Jesus's nurturing breast
and nest in Trapnel's prophecy of the post-apocalyptic future, Eve's comforting comic coda points to the aftermath of tragedy, to the hope, renewal, and transformation that are associated with the new mother after
the work of mourning that tragedy and Adam, as history's tragic subject,
perform. The long tragic drama of history in Books 11 and 12 is followed
by a brief domestic comedy that turns the household-and-cosmic world
upside-down. Woman as wife and mother, normatively revered for her
silence, is on top, while man as biological father is rendered speechless,
fruitless, and futureless. Adam gains only nominal governorship of the
reformed domestic space of the national future. "Now lead on," (12: 614)
Eve commands her husband, who is her silent partner in the new order:
he "answered not" (12: 625).

To be sure, this innovative revision of marriage and the family is de-
ferred until the End of Days. As Michael reminds Eve, in the fallen

---

[38] David Loewenstein, *Milton and the Drama of History: Historical Vision, Iconoclasm,
and the Literary Imagination* (Cambridge: Cambridge University Press, 1990), p. 92.

present moment, marriage subjects wife to husband: "Thy husband, him to follow thou art bound" (11: 291). In the end, however, the epic asks us to place comedy over tragedy, the low form over the high one—inversions consistent with the poem's anti-hierarchical, anti-traditional politics and theology. Eve's speech demonstrates how postscript becomes prophecy, how after becomes anterior in the new order. Her speech not only enacts what Norbrook describes as "the elevation of the familial narrative to the dignity of more conventional public affairs," but it also elevates the spiritual mother over the biological father.[39] Through these inversions of the paternal and the maternal, the public realm and the domestic sphere, Milton feminizes the salvific translation of the old family/nation into the new one.

Although the spiritual mother and the comic future ultimately prevail, they, however, do not completely erase the unrelentingly tragic history and compromised biological fatherhood that are revealed to and subjectively experienced by Adam. After much hermeneutical labor, Adam finally accepts that his Edenic role as a naturally procreative patriarch is compromised, as is the once joyful imperative to bear biological offspring. The powerful "bond of nature" (9: 956), which intimately links parent to child, generation to generation, and Adam to Eve, is no longer vital in the post-dynastic family-nation of Christian love and liberty. Choosing the fallen Eve over the transcendent deity, Adam in Book 9 preserves the "bond of nature" (9: 956); he is not yet aware that the Fall not only drains this bond of its prelapsarian meaning and value but it also transforms natural intimacy into unnatural enthrallment, as the slippage between "bond" and "bondage" underscores. Milton's depiction of the nature-bound, uxorious Adam in Book 9 implies contrast with the Jesus of Matt. 10:35–37, who states that he is "come to set a man at variance against his father, and the daughter against her mother...He that loveth father or mother more than me: and he that loveth son and daughter more than me is not worthy of me." Acting both on the paradisal principle that "God and Nature bid the same" (6: 176) and in accordance with the Hebraic monism that governs the prelapsarian universe, Adam refuses to be set "at variance" with Eve, despite her fallenness. But, to his horror, he learns that the monism of Paradise is perverted by the Fall. Alienated from God, Nature turns against her own offspring. Rather than increase and multiply, she gives birth to her children in the spring only to kill them off in the winter. That "our *Nourse* should overlay us, and *Ayre,* that nourishes us, should destroy us" is unthinkable, writes John Donne in *Devotions on Emergent*

---

[39] David Norbrook, *Writing the English Republic*, p. 482.

Occasions, XII,[40] but it is also a perpetually reiterated, natural matter of fact. In his postlapsarian soliloquy, Adam reluctantly recognizes that natural childbirth, hitherto the occasion for rejoicing, becomes a source of pain and sorrow. While pronounced at the Creation as good, after the Fall, natural fertility and reproductive prolificness broaden the scope of mortality, turning God's first law to be fruitful and multiply into a death sentence, endlessly reiterating Sin's unnatural birthing of Death. With the postlapsarian Adam, we grieve for the loss of increase or natural offspring, for the closeness of genealogical connection between parent and child, and for the pleasure of erotic love that knits individuals both physically and spiritually together into holy union, sanctified by marriage. "we are one,/One flesh," Adam says to the fallen Eve (9: 958–59)—a pronouncement that spells his downfall and the cosmic Fall as well. The repetition through the line break of "one/One" ironically splits One into two, evoking the rift in Adam's natural union with his spouse. This may be why Milton's epic devotes far more space after the Fall to Adam's breakdown into despair in Book 10 and his protracted initiation into new-modeled happiness in Books 11 and 12 than to Eve's unmediated vision of history's comic ending in her final speech.

Although brief and imprecise, Eve's final speech nevertheless notably contains the plot lines for Milton's program for personal and collective resurrection. As Achsah Guibbory observes, "*Paradise Lost* replaces the church with the domestic sphere as the center of religious life," but "the Fall shows the failure of even Edenic marriage [and domesticity] to be a perfect site of true worship."[41] Eve's final speech compensates for this failure by offering a script for reforming the fallen domestic sphere: for renewing its potential as "a perfect site of worship" and of regenerative community. In *Paradise Regained*, the private, maternal-centered household marks the beginning of Jesus's sojourn in the desert, and the place to which he quietly returns after his triumphant resistance to Satan's alluring temptations: Jesus "unobserved/Home to his mother's house private returned" (*PR*, 4: 638–9).[42] The turn away from the public and male and to the private and female in *Paradise Regained* reflects the enhanced meaning and value that Restoration dissenters accorded to the "mother's house private" as a new center for carrying out the unfinished work of Reformation.[43] In the

---

[40] John Donne, *The Complete Poetry and Selected Prose of John Donne*, ed. Charles M. Coffin (New York: Modern Library, 1952), p. 432.

[41] Guibbory, *Ceremony and Community*, p. 217.

[42] See Janel Mueller, "Dominion as Domesticity: Milton's Imperial God and the Experience of History," in *Milton and the Imperial Vision*, ed. Balachandra Rajan and Elizabeth Sauer (Pittsburgh: Duquesne University Press, 1999), p. 45.

[43] See Guibbory, *Ceremony and Community*, p. 190.

context of Restoration politics, Eve's visionary speech looks forward to the re-emergence of the private, domestic realm as a counter-cultural arena that could nurture dissent, which had been suppressed by Charles II's return to the English throne in 1660 and the Uniformity Act of 1662.

In *Paradise Lost*, however, Milton's comic, maternal, and nurturing vision of the emancipation of the Christian spirit and the revival of religious and political dissent is shadowed by the militaristic power of Michael's "sword": his post-Pauline dualism, which dominates the construction of postlapsarian experience in the epic. John Rogers brilliantly analyzes the surprising authoritarianism of Books 11 and 12 and its marked difference from the liberal Vitalism and materialist monism promulgated elsewhere in the epic.[44] More still needs to be said, however, about how monism of the paradisal books and the dualism of the last two books relate to the free "spirit of unity" or true restoration associated with Eve, in her final speech. Daniel Boyarin's analysis of the cultural and political implications of post-Pauline Christianity can help us to understand the terrible price that Milton never forgets must be paid in order to acquire Christian redemption with its politically subversive, comic inversions of male and female, public and private, and body and spirit. As already noted, Books 11 and 12 are preoccupied with Pauline distinctions between flesh and faith, law and gospel. As Boyarin argues, Paul's dualism allows for "a disavowal of sexuality and procreation, of the importance of filiation and genealogy, and of the concrete, historical sense of scripture, of, indeed, historical memory itself."[45] In order for Milton to make the new post-natural national future really Real (i.e. to undermine monarchy's legitimating corporealist analogy between the natural and social orders, and to rewrite Israel in spiritual terms), Milton must reduce the concreteness of the Hebraic past and the biblical Israel to dead-letter status. As he maintains in *Tetrachordon*: "Christ having cancell'd the hand writing of ordinances, which was against us, *Coloss.* 2.14, and interpreted the fulfilling all through charity, hath in that respect set us over the law, in the free custody of his love, and left us victorious under the guidance of his living Spirit, not under the dead letter" (*CPW*, 2: 587–8). This is a price that the poet ultimately proves unable to pay.

Paradise Lost thus takes the shift from dynastic to post-dynastic community and maps it onto Paul's, and later Augustine's, dualist separation of carnal Israel from spiritual Israel. Milton associates the former with the old, natural father and the latter with the new, spiritually nurturing

---

[44] John Rogers, *The Matter of Revolution: Science, Poetry, and Politics in the Age of Milton*, pp. 161–5.
[45] Daniel Boyarin, *Carnal Israel: Reading Sex in Talmudic Culture* (Berkeley: University of California Press, 1993), p. 6.

mother. Affirming Paul, Augustine writes, *"Behold Israel according to the flesh"* (I Cor. 10:18) in his *Tractatus adversus Judaeos.* "This we know to be the carnal Israel; but the Jews do not grasp this meaning, and as a result they prove themselves to be indisputably carnal."[46] For Augustine, the carnality of the Israel of Hebrew scripture and the carnality of the Jews' understanding consign both nation and people "forever to the realm of the flesh."[47] Adam, in Books 11 and 12, tries and fails time and again to repress the carnal or literal understanding that will confine him "forever to the realm of the flesh." In the end, through Michael's pedagogical persistence with his less-than-promising student, Adam does achieve the requisite spiritual insight. He embraces the redemptive implications of allegory and abstraction over a concrete sense of history, and he elevates the spirit over the body, the spiritual mother over the biological father. As Eve's prophetic speech at the end of Book 12 underscores, the alternative, woman-on-top, upside-down world of the redeemed future will be ushered in by the new spiritually nurturing mother associated with the milk of the gospel, not the old punitive, natural father associated with Mosaic Law, and its many prohibitions. The reformed family-nation of the redeemed future will consist exclusively of elect sons and reformed mothers, as Mary's and Jesus's exemplary household, "the mother's house private," in *Paradise Regained* (4: 639) suggests.

Unlike Adam, Eve smoothly crosses the divide between the old and the new dispensations. Hence, ultimately, she, not he, heralds the restoration of the old Israel in new spiritual terms. The literalist Adam and carnal Israel are relegated to the outmoded, embodied, and particularized past, to be superseded by the spiritually vital reign of the Son. In contrast to the diminished and silent Adamic progenitor, the Son represents the nourishing, spiritual-maternal, plenitude of the Word, which inscribes the new salvific era. However, as already intimated, the Pauline dualism that dominates the last two books conflicts with the poet's fervent investment in monism elsewhere in the epic and other texts (such as the divorce tracts). Milton, try as he might, cannot quite secure his triumphant, conversionist, female vision of the national future. Bounded and expansionary, secessionist and supersessionist, Milton's modern vision of the reformed nation (as discussed in my Introduction), with its foundational contradictions, must simultaneously suppress *and* retain the monism of Hebrew scripture and post-biblical Judaism's emphasis on the body, particularity, and separateness. Milton's modern vision of the reformed nation, in short,

[46] Augustine, *Tractatus adversus Judaeos*, trans. M. Ligouri, in *Fathers of the Church*, vol. 27 (Washington, DC: Catholic University of America Press, 1955), pp. 367–44.

[47] Cited in Boyarin, p. 1.

manifests his ambivalent embrace and rejection of the Hebraic/Judaic precedent. It is this ambivalence that we consider next.

TRAGIC FATHERS/COMIC MOTHERS

Tragedy enters the epic at the beginning of Book 9, with the narrator's announcement that "I now must change/Those notes to tragic" (5–6). The notes to be changed are the sweet harmonies of Adam's contrapuntal dialogue with Raphael in Books 5 through 8. Not unlike *The Symposium*, Adam and Raphael's philosophical conversation takes place over a banquet: Eve's pristine "Rural repast" (9: 4) of "savory fruits, of taste to please/True appetite," accompanied by "nectarous draughts… from milky stream, /Berry or grape" (5: 304–7), a perfect mixture of food and/for thought, corporeal and intellectual nourishment.[48] Rather than Platonic dualism, however, Adam and Raphael's Edenic symposium—a well-tempered and sober meeting of human and angelic minds—provides the poem's most extended celebration of the Hebraic monism that governs the Edenic cosmos. This is a hermeneutic that justifies a natural one-world vision of nationhood, soon to be made obsolete ("change [d]… to tragic" [9.5–6]) by the Fall.

Not simply a footnote to Adam's and Raphael's concerns, Eve's meal is an integral part of their conversation. The collision of culinary and national appetites in Milton's account of Eve's "Rural repast" (9: 4) suggestively equates food preparation with good government in a manner shared by seventeenth-century texts on husbandry, diet, medicine, and, as we have seen, maternal nurture and wet-nursing. For example, *The Court and Kitchen of Elizabeth*, a 1664 anonymously published mock-cookbook (attributed to Thomas Milbourne), offers a royalist send-up of the housekeeping practices of Elizabeth Cromwell and her husband's failure to keep his culinary, and governmental, appetites and affairs in order:

> One day, as the Protector was private at dinner, he called for an orange to a loin of veal to which he used no other sauce, and urging the same command was answered by his wife that "Oranges were oranges now, and crab oranges would cost a groat and for her part she never intended to give it"; and it was presently whispered that sure her Highness was never the adviser of the *Spanish* War, and that his Highness should have done well to consult his

---

[48] Barbara Lewalski describes Adam's conversation with Raphael about the problem of his passion for Eve as modeled on Plato's *Symposium* and its Neo-Platonist versions by Ficino, Leone Ebreo, and Castiglione in *Paradise Lost and the Rhetoric of Literary Forms* (Princeton: Princeton University Press, 1985), pp. 214–15.

Digestion, before his hasty and inordinate appetite of Dominion and Riches in the *West Indies*.[49]

The Cromwell household is satirically depicted as the world-turned-upside-down desired by mid-century regicides, which turns out to be nothing but a monarchy manqué: her Highness surpasses his Highness. Having deposed and executed a king, Cromwell is unseated from the throne of his own household dominion, undone by his expensive tastes and unreasonable commands at the dinner table. His exorbitant culinary appetites are equated, in turn, with "his hasty and inordinate appetite" for imperial conquest in the Western Design. Undisciplined, overreaching, and intemperate both at and outside of his private table, the Lord Protector is an unruly female figure in this farcical domestic scene. His private and public excessiveness are kept in check by his wife's governorship of the household and her practical concerns with the cost of oranges and good digestion.[50]

Like *The Court and Kitchen of Elizabeth*, but from a republican vantage point, Milton's account of Eve's midday meal makes the English dinner table into a highly contested site in which the competing ideals of royalism and republicanism fight for supremacy. Of special interest is Milton's suggestion that, before the Fall, a natural vision of the national body politic is compatible with an anti-hierarchical framework for organizing social and domestic relations—this is in ironic contrast to the fallen corporealism through which dynastic kingship legitimated its top-down power and authority. Equally notable is that the nurturing Eve forms a crucial component of this potentially fluid national order. With a "milky stream," Eve "crown[s]" the "flowing cups" (5: 445, 444) she serves up to Raphael and her husband. This crowning gesture not only ennobles the "milky" nurture she provides, but it also makes labile the degree-and-gender distinctions that otherwise hierarchically stratify Eve, Adam, and their angelic guest, fluently linking her "lowliness majestic" (8: 42) to the domestic sovereignty that she honors in and confers upon her husband and Raphael through her "crown[ing]...cups" (5: 445, 444). Woman, man, and angel congregate and commune at the same table, where "Eve/Ministered naked"

[49] *The Court and Kitchen of Elizabeth* (London: Thomas Milbourne, 1664), p. 39.

[50] Gillespie provides stimulating commentary on Milbourne's depiction of Elizabeth Cromwell in "Elizabeth Cromwell's Kitchen Court: Republicanism and the Consort," *Genders*, 33 (2001): 29–30. Knoppers contrasts the depiction of Elizabeth Cromwell as a country bumpkin and stingy foreigner in "Elizabeth Cromwell's Kitchen" with Henrietta Maria in the "Queens Closet Opened," transformed from detested French Catholic Queen into a good English housewife in "Opening the Queen's Closet: Henrietta Maria, Elizabeth Cromwell, and the Politics of Cookery," *Renaissance Quarterly*, 60.2 (Summer 2007): 464–99.

(5: 442–3). She at once serves and governs without ceremony, fanfare, title, or in a manner consistent with the poem's Puritan and republican antipathies to kingship, theater, and Catholicism. The naked Eve is also unmarked by class, economic, and other external and particularizing inscriptions of rank and degree, not unlike the nursing mother of the guidebooks. As Norbrook observes, Eve's naked ministry coheres with the poem's "republican delight... in stripping away false customs"; her nudity also underscores "how deliberately provocative Milton was being in glorifying his characters' sexuality."[51] Linking human sexuality to social and divine justice, the naked Eve and her well-tempered table encapsulate the monism that organizes prelapsarian experience, in which body and spirit, nurture and nation, Nature and God, bid the same.

Eve's meal serves up a rich poetics, spelling out the primal but soon-to-be-doomed union among maternal nurture, Mother Nature, and the promised natural, nurturing nation written in Eve's milk and fruits.[52] This pristine food and drink are raw and neither spiced nor sweetened, devoid of historical, geographical, or cultural inflection, reflecting the fundamental, virginal state of Mother Nature and divine creation. Eve's meal is as unadorned and naked as is Eve herself. Although "the poetics of spice," to borrow Timothy Morton's apt phrase, is beyond the scope of this chapter, I would nevertheless like to point out that after the Fall, spice and sugar, like clothing, function in Milton's epic as shameful mercantile signs of imperial ambition.[53] Before the Fall, sugar and spice flourish in Eden's "blissful field" (5: 292): the source of "enormous bliss" (5: 297) and balm to both the body and spirit. As he approaches Adam and Eve, Raphael walks through "the spicy forest" (5: 298) as Eve prepares "savory fruits" (5: 304) for his dinner. After the Fall, however, sugar and spice become comestible markers of monarchical and prelatical decadence and despotism, and of ornamentalism and orientalism. Sensory, but decidedly non-nurturing enticements, sugar and spice both mask and make falsely palatable humankind's rapacious imperial appetites. These appetites, which are linked to Hell's bottomless maw and Death's insatiable hunger, first surface in Book 10. The Fall of Nature results almost immediately in blood-thirst and meat-eating, whereas the consumption of blood and flesh, or the body, which is, for Milton, falsely dignified after the Fall

---

[51] David Norbrook, *Writing the English Republic*, p. 481.

[52] Diane McColley, *A Gust for Paradise: Milton's Eden and the Visual Arts* (Urbana: University of Illinois Press, 1993), p. 133. McColley describes "Eve's composing of food," as "a trope for poetry."

[53] See Timothy Morton, *The Poetics of Spice: Romantic Consumerism and the Exotic* (Cambridge: Cambridge University Press, 2000) for a richly nuanced reading of Eve's midday meal in the context of early modern trade with the Orient and the semiotics of spice.

through the taking of the sacraments, is nowhere to be found in the prelapsarian cosmos. For Milton, the divinely sanctioned Edenic vision of natural empire as inscribed by the global reach of the fruits and the milky drink that Eve heaps upon her ideologically charged dinner table cannot be recuperated after the Fall, especially not by the intemperate amassing of false material goods, such as ornamental/oriental spice, sugar, silk, through trade or colonization. Such evil empire-building, epitomized for Milton by Catholic Spain and Portugal, the Stuart kings, and Cromwell's ill-fated Western Design, is modeled in the poem by Satan's vengeful incursions into the new world of Earth.[54] Milton compares Satan in conquistadorial flight to the new world of Earth to a fleet of merchant ships carrying spices back to England from the Moluccas or Spice Islands, "the isles/Of Ternate and Tidore, whence merchants bring/Their spicy drugs, they on the trading flood/Through Ethopian to the Cape/Ply" (2: 638–43). The Edenic promise of a universal natural nationhood after the Fall can be fulfilled only when translated into unspiced, naked, and spiritually virginal terms and implemented through such immaterial means as plain-styled preaching and Puritan evangelism. Gospel, the unadorned milk of the Word, as we have already seen, will be humankind's only true form of sustenance.

Idealized by the guidebooks, natural maternal nursing, by contrast, finds perverse expression in Milton's postlapsarian universe. Fallen Mother Nature's charge is to nourish Death, her surrogate son: to provide him with a never-ebbing flow of corroding bodies and souls upon which to feast—"Food for so foul a monster" (10: 986). She leaves her natural offspring to fight amongst one another for sustenance. The scarcity that sets in after the Fall transforms Paradise into a riotous realm of carnivorous hunters fighting over limited resources. The translation of the Garden into a battlefield is one of the first visible manifestations of the Fall: "Beast now with beast gan war, and fowl with fowl/And fish with fish; to graze the herb all leaving,/Devoured each other" (10: 710–12). This Hobbesian depiction of the state of nature anticipates the characterization of Nimrod as "A mighty hunter" (12: 33). Through Nimrod, the fallen urge to hunt is tied directly to the Tower of Babel, "whose top may reach to Heav'n" (12: 44) and to "empire tyrannous" (12: 32). Meat-eating, hunting, tyranny, and the pseudo-sublime prospect of empire all stem from the same perverse point of origin in Mother Nature: "Authority usurped, from God not giv'n" (12: 66) in the form of idolatry, tyranny, and the tyranny of

---

[54] J. Martin Evans, in *Milton's Imperial Epic: "Paradise Lost" and the Discourse of Colonialism* (Ithaca: Cornell University Press, 1996), argues that *Paradise Lost* registers Milton's ambivalences about European colonization of the New World.

idols. Linked to Hell by the bridge built by Sin and Death, postlapsarian Earth recalls Hell's "universe of death" (2: 622): "Where all life dies, death lives, and nature breeds/Perverse, all monstrous, all prodigious things/ Abominable, inutterable, and worse" (2: 624–6). Rather than the maternal source of biological life, fallen Nature profusely issues perverse and abominable artifacts—idols, pagan deities, and tyrants.

The perversion of Mother Nature after the Fall permeates Adam's meditations on death (and mothers) in Book 10. He associates the multiplying of sorrow after the Fall with the multiplying of progeny. "Posterity stands cursed," he laments; "fair patrimony/That I must leave ye, sons; O were I able/To waste it all myself, and leave ye none!/So disinherited how would ye bless/Me now your curse!" (10: 818–22). Rather than experience death as the return to Nature's and his "mother's lap" (10: 778), Adam wonders if his immortal soul will consign him to a painful, never-ending state of suspended animation: "who knows/But I shall die a living death?" (10: 787–8). He knows that the body is subject to death, but he thinks that the spirit is immortal: "the spirit of man/ Which God inspired, cannot perish/With this corporeal clod" (10: 784–7). Just as the Fall had divided him from Eve in those tense, internally conflicted moments in Book 9 just before he joins his spouse in mortal sin, so more generally in Book 10 does it set spirit and the matter, soul and body, at variance. Once again, Adam refuses to break the "bond of nature" (9: 956). Forgetting Raphael's eschatological conjecture in Book 5 that "Your bodies may at last turn to spirit,/Improved by tract of time" (5: 497–98), he cannot bring himself to turn from his fallen Mother Nature, or from that which is begot by "natural necessity"(10: 765) to what the deity makes "of choice his own" (10: 766). Adam longs to die a natural death in his mother's lap, but this desire tragically leads his mind only to a painful, conceptual dead end: "Will he draw out," Adam miserably wonders, "For anger's sake, finite to infinite/In punished man" (10: 801–3). This "Strange contradiction" (10: 799) plunges him into an "abyss of fears/And horrors" (10: 842–3).

Adam's suspended state between the old era and the new one coincides in theological terms with what Rosenblatt describes as "the state of misery that follows innocence and precedes grace."[55] Although he will learn to find new joy in Eve's role as spiritual mother of the promised seed, he is unable as yet to foresee history's comic conclusion in the fullness of time at the End of Days. After nature and law, but before the spirit and gospel, Adam cannot comprehend the double implications of his meditations on life and death. He does not yet know that the death sentence he now hears

[55] Rosenblatt, *Torah and Law in Paradise Lost*, p. 205.

in God's life-affirming command will be reversed when he gains what Edward W. Tayler describes as "the awareness of double duration." This split temporal consciousness surfaces in Milton's epic through the proleptic formal pattern of tragic "anticipation and fulfillment," achieved through ironic reversal. The "awareness of double duration" supplies new meaning and value to fallen human events by revaluing these temporal (and temporary) failures in relation to Christian history's salvific end point, to be ushered in by the Apocalypse.[56] In Books 11 and 12, through Michael's lessons on typological allegory, Adam learns that this proleptic tragic pattern is played out in human history culminating in a happy end—and that this end is closely associated with the new mother or second Eve. In his soliloquy, however, he remains divided against himself: he is the split subject of the epic's Hebrew tragedy.

The typically strict separation of mourning from rejoicing in Hebrew scripture further illuminates the larger epochal shift that Milton dramatizes through Adam's soliloquy. Adam's failure to distinguish between misery and happiness, death and life is notable because, this kind of conflation of mourning and rejoicing almost never occurs in Hebraic scripture, with the exception of Amos 8:3 and Jer. 41:5.[57] As Saul Olyan observes, "mourning and cultic rejoicing are typically represented in biblical texts as antithetical and incompatible set of ritual behavior," but in Jer. 41:5 ("Eighty men came from Shechem, Shiloh, and Samaria, their beards shaved, their garments torn, and their bodies gashed, carrying meal offerings and frankincense to present at the House of the Lord") where, in the face of the destroyed temple, pilgrims bear offerings of pilgrimage in the guises of mourners, these normally incompatible rites are uncharacteristically fused "into a single ritual constellation." This can only mean "that the distinctions between the contrasting states of cultic rejoicing and mourning have completely broken down, and therefore the ritual order has collapsed." The fusion of mourning and rejoicing in Jer 41:5 epitomizes "the way members of the community might speak about disaster *in the most profound and striking terms possible.*" Adam's fusion of misery and happiness echoes Jer 41:5, by speaking "*in the most profound and striking terms possible*" of the collapse of the "separate, sanctified, cultic sphere," which is Eden. Before the Fall, Edenic nature is the sanctified sphere of ritual order, as expressed by Adam's and Eve's spontaneous prayers and the mysterious rites of marriage that they observe naturally in their bower of bliss. The Fall renders ritual worship unnatural.

---

[56] Edward W. Tayler, *Milton's Poetry: Its Development in Time* (Pittsburgh: Duquesne University Press, 1979), pp. 132–5.

[57] Saul M.Olyan, "'They Shall Wail the Songs of the Temple': Sanctioned Mourning in Biblical Cultic Settings," p. 11.

Yet, because Adam is situated in the unstable space between the old and new dispensations, his fusion of misery and happiness, mourning and rejoicing, evokes not only the breakdown of the old ritual order in Hebraic terms, but also the new creation heralded by the Pauline doctrine that informs the last two books of Milton's epic. In Galatians 3:28, Paul cancels the differences between the old and new orders by negating embodied forms of distinction, such as ethnic, gender, and social difference: "There is neither Jew or Greek, there is neither slave or free, there is neither male and female; for you are all one in Christ Jesus." Galatians 3:28 echoes and reverses Genesis 1:27: those in Christ have entered a new creation where former creative distinctions such as light and darkness, day and night, mourning and rejoicing, death and life, are replaced by a new recreative unity. If the collapsing of distinction spells irreversible loss in the most poignant terms possible within Hebraic scripture, in the Galatians verse the same conflation of difference powerfully proclaims new creative union and community.

Galatians 3:28 also informs the consoling, chiasmic Christian doctrine of life as death and death as life, which finds repeated expression in early modern English religious discourse. Countless sermons, verses, and prayers conditioned believers to await a happier life in the after-life. "The day of our death is better than the day of our birth," preached Thomas Playfere in 1595. "For when we are born we are mortal, but when we are dead we are immortal. And we are alive in the womb to die in the world, but we are dead in the grave to live in heaven."[58] The dying man in Samuel Hieron's *Help Unto Devotion* is assured that an angel will come "to lead me out of prison into liberty … from this vale of tears into eternal bliss."[59] John Donne vividly renders this commonplace pulpit conceit—that the end of life is inseparable from the beginning of life—in one of his sermons from 1630: "We have a winding sheet in our mother's womb and we come into the world wound up in that winding sheet, for we come to seek a grave."[60] Dualism is the governing principle behind these consoling sentiments. As David Cressy points out, "Central to these discussions was the theory of the separation of body and soul."[61] As yet untutored in Pauline doctrine, Adam cannot convert his Hebrew

---

[58] Thomas Playfere, *The Meane in Mourning* (STC, 20017, 1597), 26, 56–60. David Cressy, *Birth, Marriage, and Death: Ritual, Religion, and the Life-Cycle in Tudor and Stuart England* (Oxford: Oxford University Press, 1999), p. 566. Playfere's sermon was reprinted in 1607 and 1616.

[59] Samuel Hieron, *A Help vnto Deuotion: Containing Certain Moulds or Forms of Prayer* (STC 13407, 1621), 260.

[60] John Donne, *The Sermons of John Donne*, ed. George R. Potter and Evelyn M. Simpson (Berkeley and Los Angeles: University of California Press, 1962), 10.233.

[61] David Cressy, *Birth, Marriage, and Death*, p. 384.

monism into dualism. He thus fails to find consolation and remains entrapped by his grief and mortal terrors.

Milton's poetic sublimity expresses in aesthetic terms Galatians 3:28's doctrinal emphasis on neither/nor, or spiritual transcendence.[62] In *Paradise Lost*, moreover, the sublime is additionally related to the new spiritual mother—a remarkable but curiously overlooked innovation. Remembering "Part of our sentence" (10: 1031), Adam associates this "more sublime" (10: 1014) something in Eve with "thy seed" (10: 1031), the promised seed, from which will spring a new generation, or, in Pauline terms, the new "children of the promise," for whom the external signs of ethnic, gender, and class difference are irrelevant. External and material differences will resolve into and create new internal, spiritual harmonies. The sublime "children of the promise" will supersede the outmoded "children of the flesh," or Adam's cursed, natural race. Lacking grace, Adam as yet does not comprehend the full meaning of the sublime-Pauline insight that he will derive from his more abstract reading of his sentence. As we shall soon see, his difficulties in dissociating the woman's seed from his own loins and biological progeny will impede his path toward sublimity and typological understanding in Books 11 and 12. Despite his interpretive limitations, however, Adam's response to Eve nevertheless emphasizes the close affinity between the sublime and the new mother.

This maternal aspect, as we recall from Chapter 3, derives from Longinus's depiction of the sublime as the nurse of liberty. Norbrook attributes the Miltonic sublimity to the republican revival of Longinus's treatise, *Peri Hypsous, On the Sublime*. Interest in the sublime and in recuperating Longinus "was found in royalist as well as parliamentary circles, but Milton and his followers gave the sublime a distinctively republican accent." Milton "had placed Longinus on his ideal syllabus in *Of Education*, and praised poetry as 'that sublime art'; it was appropriate that the first English translation of Longinus should have been composed by [Milton's] young disciple, John Hall." (Hall's translation was published in 1652.)[63] We can gain even deeper insight if we read the maternal aspect that Milton associates with the sublime and the spirit of the free nation in relation to his Pauline perceptions of Christian liberty. Despite their

---

[62] Important commentary on the Miltonic sublime can be found in Nigel Smith, *Literature and Revolution in England, 1640–1660* (New Haven: Yale University Press, 1997), p. 125. Smith argues that the category-breaking energies of the Miltonic sublime are related to the free nation's anticustomary spiritual power to reconcile opposites. David Norbrook associates Milton's use of the sublime in *Areopagitica* with the poet's desire to release the spirit of the free nation from the diseased body politic of Stuart monarchy, but Norbrook nevertheless praises "Milton's monism, his rejection of a sharp split between spirit and matter," *Writing the English Republic*, p. 137.

[63] David Norbrook, *Writing the English Republic*, pp. 137–9.

apparent disparities, Milton's anti-corporeal, Pauline theology and his iconoclastic Longinian sublimity mutually shape his proto-modern, post-dynastic vision of the nation/family, including the central role that the new mother plays in ushering in the reformed national future. Paul's devaluation of the body legitimates Milton's devaluation of nature, the biological father, and the dynastic nation. The poet's Paulinism clears new conceptual space for his sublime, spiritual-maternal ideal of anti-dynastic community, which negates and purifies blood-and-soil justifications for kingship and the received social conventions, religious rituals, and set forms that validate dynastic government. Upon this *tabula rasa*, he inscribes his apocalyptic vision of collective harmony, a vision indebted to the sublime category-bursting ("There is neither Jew nor Greek") unity and community celebrated in Galatians 3:28. Instead of the old paternal bonds of blood and kinship, new maternal bonds of love and the spirit organize the reformed nation.

In poetic terms, the sublime's anticustomary harmonies (*discordia concors*) most closely connect Milton's Pauline vision of new creation and community to the reformed trope of the new mother. Pushing against the smooth language and balance that characterize Isocrates's eloquence—a symmetry recreated by the rhyming heroic couplets of Royalist verse—Longinus associates the sublime with asymmetrical harmonies: "things...widely different are here by a strange artifice brought together"; the very order seems to be disorderly. The "strange artifice" that produces the sublime union of "things...widely different" forms the aesthetic equivalent of the spiritual reformation of nature: the prerequisite needed to harmonize and redeem fallen human community in Pauline terms or as subject neither to decay, corruption, death, nor any of the embodied particularities of native or ethnic nationhood ("Jew or Greek"), class ("bound or free"), or gender ("male or female") that limit and hence impair the unforced, anti-customary governance of free will. Whereas Sin, the bad mother, joins "With secret amity things of like kind," through her "stupendous bridge" (10: 351) between Hell and Earth, the regenerative *discordia concors* of the sublime, the maternal nurse of liberty, recreates unity and community in the spiritual terms celebrated in Galatians 3:28. And, whereas the discordant harmonies of sublime have a republican accent, Sin's uniting of like with like recalls the convention-bound, rhyming symmetries of the heroic couplets featured in royalist verse: "the Invention of a barbarous Age," as Milton writes of rhyme in his preface on "The Verse."

In *Areopagitica*, Milton allegorizes his maternal, sublime-Pauline ideal of community when he depicts the newborn English republic as an arisen female Samson—a flight of prose-poetry that pushes the traditional organic analogy between the healthy natural body and the English body-politic to breaking point. This sublime moment of rupture releases the newborn spirit of the nation from the corrupt, old body politic, which Adam's lament subjectively re-creates. Bereft of grace, Adam experiences this joyful rebirth as a painful wound: as human tragedy not as divine comedy. Adam dramatizes how the supersessionist transformation of the old Israel into the new one inwardly afflicts the tragic postlapsarian male subject, not to mention the generic and aesthetic forms through which these transformative subjective conflicts find expression. In the end, the soliloquized despair of Adam's Old Testament jeremiad is converted into his sublime-Pauline "joy and wonder" (12: 468). His emotional reformation begins at the end of Book 11, with his conceptual breakthrough from literal to allegorical interpretation. Adam understands the typological significance of the olive leaf born by the ascending dove: that it foreshadows the end of the old Hebraic dispensation and the beginning of the New Covenant of the gospels. In his earlier lament, however, he lacks the grace necessary to experience the transcendent sublimity of redemption. He is mired in the miserable/happy temporal space between the Old and New Testaments, between the tragic patriarchal drama of history and the apocalyptic maternal comedy of redemption. The various polarities that Adam piles up in his Jeremiad-like lament poignantly express the impossibility of mourning the irreversible collapse of the sanctified Edenic order and the irreparable rupture of God and Nature after the Fall. At the same time, they point happily to the death of kingship and to the post-dynastic, sublime-Pauline community of the spirit into which Adam will be initiated by Michael in Books 11 and 12. Adam's lament models the tragic, Hebraic language of exile, wandering, and error, into which the lost perfection of Eden's monistic discourse is tragically translated.[64] For those who have received gospel and grace, however, the collapse of distinction that characterizes Adam's tragic lament not only miserably spells out the end of the old order and "Israel according to the flesh" (1 Cor. 10:18), but it also happily creates the *discordia concors* that is the higher harmony of the sublime new maternal order: the spiritual Israel, celebrated by Galatians 3:28 and "*the unity of spirit*" that obtains among the free citizenry of the godly nation.

These new sublime harmonies find their way to the lamenting Adam through Eve, anticipating the evangelizing spirit through which gospel

---

[64] In Ps. 107:40, God "causeth them to wander in the wilderness, where there is no way"; in Job 12:24, He "causeth them to wander in a wilderness where there is no way." David Quint associates wandering with political liberty and linearity with political conformity in *Epic and Empire: Politics and Generic Form from Virgil to Milton* (Princeton: Princeton University Press), pp. 43–4.

will be disseminated. Devoid of grace, the lamenting Adam mistakenly repudiates Eve's belatedness. He is unable to recognize that, rather than an after-thought of the old dispensation, she is the harbinger of the new one:

> O why did God
> Creator wise, that peopled highest Heav'n
> With spirits masculine, create at last
> This novelty on Earth, this fair defect
> Of nature, and not fill the world at once
> With men as angels without feminine,
> Or find some other way to generate
> Mankind?   (10: 888–95)

Adam's anti-female invective reflects the enduring tradition of woman-hating literature, of which Joseph Swetman's popular *The Arraignment of Lewd, idle, froward, and unconstant women* is an especially vivid example. Here, as elsewhere in the postlapsarian books, the central question for Adam concerns biological reproduction and progeny, or how naturally to "fill the world." Disgusted by the prospect of naturally procreating with "this fair defect/Of nature" (10: 891–2), Adam nevertheless refuses to give up on biological reproduction until almost the very end of the epic, even after he learns in Book 12 that the children of Abraham's promise will supersede the children of Abraham's loins. He also fails to recognize that the spiritual maternal line, from which he is excluded, carries the promise of redemption. Jesus is the "son of Mary second Eve" (10: 183). Eve, "this novelty on Earth," created "at last" (10: 890–1) ultimately shall displace and de-center all that naturally precedes her, including, or especially, Adam's natural patrimony.[65] In political terms, her reformed maternal role allegorizes the ultimate triumph of the officially despised and politically dispossessed, coded feminine. The topical relevance to Milton's own outlaw and outcast status and to all those disenfranchised by the restored Stuart monarchy would not have been missed by the poet's fit readers. Eve's final speech offers hope to the defeated: "though all by me is lost/.../ By me...shall all restore" (12: 621–23).

As ever, however, alongside Eve's Christian comedy of dissent and sub-version is "the Hebraic tragedy of Torah degraded into law and of redemption purchased at a terrible price," to repeat Rosemblatt's poignant observation.[66] Eve's exemplary conversion from defeat to triumph, from the victim to the agent of providential history, parallels the antitheses of

---

[65] See Diane McColley, "Subsequent or Precedent? Eve as Milton's Defense of Poesie," *Milton Quarterly*, 20 (1986): 132–6.
[66] Jason Rosenblatt, *Torah and Law in Paradise Lost*, p. ix.

the gospels. But, these triumphant Christian reversals also tragically reduce the concreteness of the Hebraic past into a sign or shadow of that which Milton's epic deems really Real, i.e. the unity of spirit that obtains amongst the free citizens of the godly nation. As Boyarin maintains, Pauline doctrine allows for "a disavowal of sexuality and procreation, of the importance of filiation and genealogy, and of the concrete, historical sense of scripture, and or, indeed, historical memory itself."[67] Milton's Pauline vision of the salvific national future thus cannot be realized without repressing the body and forgetting the embodied Hebraic past. Adam must make this sacrifice so that the post-dynastic nation inspired by Galatians 23 can be realized in spiritual terms. Just as his new insight into Eve's sublimity in Book 10 leads him out of the mental labyrinth that he creates in his lament, so Adam's Pauline embrace of the spirit over the body, of the allegorical over the literal, gives him a way to exit history's reiterative and relentlessly tragic drama and enter into the redemptive, comic future of the world-turned-upside-down. The regenerative future belongs to the materially dispossessed, politically disenfranchised, and, more specifically, the Protestant diaspora of those exiled (spiritually and physically) by the Restoration of Stuart kingship from their national homeland. Paul's denigration of the body and nature grants those without blood-and-soil measures of social/cultural distinction new value, honor, and nobility in abstract or spiritual terms. In the new era and the second Israel, spiritually regenerate Others (and [M]others) will prevail over their ungodly masters—the hyper-masculine Nimrods, who tyrannize the old world.

But while Adam's disavowal of the body, sexuality, and biological descendants rewards him with a happy, sublime-Pauline prospect of the comic future, his newfound joy does not completely cancel out the protracted tragedy of history that dominates the epic's last books. As we shall see next, despite Milton's ideological and theological investments in Pauline doctrine, the poet cannot completely do away with the body, the embodied past, or the matter of memory itself.

FILLING THE FUTURE AND ANSWERING NOT

Adam's ambivalence about the proper relationship between body and spirit, old fathers and new mothers, finds especially clear expression near the end of Book 12. Michael pauses after narrating the gospel tale of Christ's miraculous birth, from which he notably omits all reference to

---

[67] Boyarin, *Carnal Israel*, p. 6.

Jesus's human father: "A virgin is his mother, but his sire/The power of the Most High" (12: 368–9). Having elicited the proper affective response from his student, "such joy/Surcharged, as had like grief been dewed in tears," Michael allows Adam "the vent of words" (12: 373–4). Under the tutelage of his "sublime" (11: 237) mentor, Adam's grief converts to joy and his discord to harmony, such that he can now arrive at and articulate a breakthrough moment of sublime insight, "Now clear I understand" (12: 376). "Now," in this instance, is a temporal marker both of the present and of eternity, of history as progressive revelation (or prophecy), from birth to rebirth, rather than as genealogical descent:

> O prophet of glad tidings, finisher
> Of utmost hope! Now clear I understand
> What oft my steadiest thoughts have searcht in vain,
> Why our great expectation should be called
> The seed of woman: Virgin Mother, hail,
> High in the love of Heav'n, yet from my loins
> Thou shalt proceed, and from thy womb the Son
> Of God Most High; so God with man unites.
> Needs must the serpent now his capital bruise
> Expect with mortal pain.
>
> (12: 375–83)

After hearing the story of Christ's nativity, Adam can at last interpret the mysterious terms in which "Jesus, son of Mary second Eve" (10: 183) sentences the serpent, that the woman's "seed shall bruise [the serpent's] head" (10: 180–1). But, in alluding to "my loins," he fails to recognize that he, the archetypal biological father, is missing from Michael's account of the Nativity. Adam disavows that the Son's identity is carried on the maternal line, and his denial impairs his perception of historical redemption. If Adam's first "Now" ("Now clear I understand"), which is a marker of both the present and of eternity, demonstrates his understanding of history as progressive revelation, his second "now" ("the serpent now with capital bruise/Expect mortal pain") confines temporality to the present moment, and reduces history to chronology, devoid of providential purpose.

Having heard the story of Christ's nativity, Adam can now translate biological procreativity and embodied life into spiritual terms: "Virgin Mother, hail" (12: 379). Adam's salutation highlights his sublime association of spiritual rebirth with the new mother. This is the insight that Michael, the "sublime" (11: 237) teacher, has patiently tried to elicit from his heretofore stubbornly literalist student. Yet, even here, Adam does not quite deliver the requisite response. As his speech makes clear, he happily embraces the primacy of spiritual motherhood, but he refuses to recognize and accept the obsolescence of his own biological patrimony and paternal line. Although Michael omits Adam from his account of the Nativity, Adam writes himself back in: "yet from my loins/Thou shalt proceed" (12: 380–1).

The fate of "my loins" (12: 380), sexuality, the body, and natural progeny thus remain the preeminent issues for Adam, and it is no accident that Milton should highlight them. Resistance to dualism in language, body, and nationhood, such as that which Adam recurrently displays, marks *the* defining point of contention and difference between rabbinic Judaism and post-Pauline Christianity in late antiquity. For Boyarin, this resistance was "at least partially owing to cultural politics."[68] The contestation around the body between rabbinic Judaism and its Hellenic competitors, including Paul, modulates into a contest between a literal versus a figurative interpretation of history, scripture, and religious practices, "but it is nevertheless, the same contest," forming "a nexus between the interpretations of sexuality and the interpretations of ethnicity." Because of their shared Hellenic background, early Christianity and rabbinic Judaism recognized the same nexus between hermeneutics and cultural politics, but each interpreted these hermeneutical and cultural–political interrelations in polar opposite ways.

How do these intermingling concerns about the allegorical and the literal, the spirit and the body, hermeneutics and cultural politics intertwine with the epic's narrative of regenerate nationhood? As already noted, *Paradise Lost* maps the Hebraic–Christian divide onto the boundary between dynastic and post-dynastic community, and between the organic and spiritual nation. In Books 11 and 12, Milton further conflates the post-dynastic nation's spiritual triumph over traditional, monarchical formulations of the body politic with the triumph of Christian allegory over Hebraic literalism. He restages the period's contest between kingship and the non-dynastic commonwealth, and between the external and internal or symbolic measures of social status and political community, as a hermeneutical contest between the body and spirit, but these contests are fundamentally the same. The unity of spirit described by Galatians 3:28 defeats the monarchical body politic; however, Milton cannot quite give up the body and monism entirely. This is because his modern view of the post-dynastic nation as both spiritual and particular depends upon the same intertwining Hebraic concepts of election, separateness, and embodiment that he otherwise represses through the radical Paulinism of the last two books.

---

[68] Daniel Boyarin, *Carnal Israel: Reading Sex in Talmudic Culture* (Berkeley: University of California Press, 1993), p. 6.

These ambivalences and ironies are acutely felt in Books 11 and 12. Adam's progress from literalism to abstraction, from in-the-flesh to in-the-spirit thinker, refracts Milton's reformist vision of national progress from dynastic government and its corporealist paradigm of nationhood into post-dynastic community or the nation in its modern, abstract, and spiritual guise. Adam's "from my loins" (12: 380) speech offers an especially suggestive example of the interlocking relations that the epic establishes between Christian allegory and modern nationalism, as well as their maternal implications. Adam, in embracing the new mother, has absorbed and begun to master Michael's lessons in the outer–inner, male–female, visible–invisible, body–soul dichotomies of allegorical reading, thus moving away from carnal thinking toward spiritual insight. But, his allusion to the foundational importance of the body and biological fatherhood ("yet from my loins/Thou shalt proceed"[12: 380–1]) suggests that he has not fully crossed (or cannot cross?) the old/new, male/female, Hebrew/Christian, natural nation/spiritual nation, birthright/merit divide.

Rather than Paul's allegorical formulation of redeemed identity, Adam, in this speech, evokes something close to the spiritualized and philosophical mode of Judaism espoused by Philo, which, while incorporating Christian allegory into Jewish thought and religious and social practice, nevertheless preserves reverence for the body. Hence, while underscoring the importance of "the inner meaning of things," Philo argues not to "repeal the law laid down for circumcising":

> It is true that receiving circumcision does indeed portray the excision of pleasure and all passions, and the putting away of the impious conceit, under the mind supposed that it was capable of begetting by its own power; but let us not on this account repeal the law laid down for circumcising. Why, we should be ignoring the sanctity of the Temple and a thousand other things, if we are going to pay heed to nothing except what is shown to us by the inner meaning of things. Nay, we should look on all these outward observances as resembling the body, and their inner meanings as resembling the soul. It follows that, exactly as we have to take thought for the body, because it is the abode of the soul, so we must pay heed to the letter of the laws.[69]

Paul's allegorical reading of the rite of circumcision makes exactly the opposite point, separating ritual circumcision and in-the-flesh thinking from the figurative circumcision of the heart and spiritual thought:

---

[69] Philo, "The Migration of Abraham," *Loeb Classics Philo* (London: Heinemann, 1932), 4.185.

For when we were still in the flesh, our sinful passions, stirred up by the law, were at work on our members to bear fruit for death. But now we are fully freed from the law, dead to that in which we lay captive. We can thus serve in the new being of the Spirit and not the old one of the letter (Romans 7:5–6).

The rabbis' response to Paul's interpretation of the rite of circumcision and its equation of the body and sexuality with death was strongly to insist upon the exclusive particularity of Judaic bodies and genealogies. As Boyarin observes, such resistance is "both the distinction of rabbinic Judaism and its limitation." At the same time, "the universalizing possibility of post-Pauline Christianity, with its spiritualizing dualism, was also obtained at an enormous cost." If the insistence in rabbinic Judaism "on corporeal genealogy and practice of tribal rites and customs produces an ethnocentric discourse, a discourse of separation and exclusiveness," then the emphasis within post-Pauline Christianity on "the allegorization, the disembodiment of those very practices produces the discourse of conversion, colonialism, the White Man's Burden, Universal Brotherhood in the body...of Christ."[70] Rabbinic Judaism and post-Pauline Christianity emerge in response to one another; their limitations (hyper-awareness of the body on the one hand, and disavowal of the body on the other hand) thus are also weirdly contiguous.

Milton's project in *Paradise Lost* is to revise the interconnections between rabbinic monism and post-Pauline dualism, such that their mutually implied limitations would cancel one another out. This new merger of the "Judeo-Christian" underwrites a vision of reformed nationhood that is at once secessionist and supersessionist, ethnocentric and expansionist, separatist and universal.[71] *Paradise Lost* assigns this consolidating task to second Eve. The new mother nourishes the common ground between the two dispensations. By extension, she also nurtures the multiple body–spirit contradictions that constitute modern nationhood: i.e. that the nation is horizontally unified in spirit by its shared principles and fellow-feeling, while, at the same time, it is vertically stratified into differing social/economic classes (as noted in my Introduction). But, Milton's epic also exposes the enormous cost of harmonizing the new nation's foundational oppositions through the spiritual mother. These include the

---

[70] See Boyarin, *Carnal Israel*, pp. 6, 233, and 235. Schlomo Sand argues against the idea that Jewish isolationism emerges in dialectical opposition to Christian expansionism in *The Invention of the Jewish People*, trans. Yael Lotan (London: Verso, 2009), Chapter 3. He argues that proselytism (with an eye toward empire-building) was a common Jewish practice in antiquity.

[71] Linda Gregerson, "Colonials Write the Nation: Spenser, Milton, and England on the Margins," in *Milton and the Imperial Vision*, p. 169.

erasure of the material history of the biblical Israel—and other concrete histories of embodied others—with all its broader colonizing implications. Milton, in the end, proves unable to purchase national redemption at this terrible price. He cannot quite forget to remember, or remember to forget, the concreteness of the past, as well as the very matter of cultural memory.

On the one hand, as Adam's Philo-esque "from my loins" (12: 380) speech suggests, Milton strives in his epic (and in *Samson Agonistes*, as we shall see in the next chapter) to restore the Hellenic context in which the emphatically separatist, body-centered discourse of rabbinic Judaism and the spiritual dualism of post-Pauline Christianity could be brought into fruitful conceptual alliance; but, at the same time, he overlays this experiment in productive Judaic-Christian, body–spirit collaboration with a supersessionist interpretation of Judeo-Christian history, which reduces the concreteness of Hebrew scripture's history of Israel to phantom status: a series of disembodied types awaiting re-presentation and spiritual fulfillment in the new era. In so doing, Milton can give up on the Judaic insistence on separatism and exclusion ("the distinction of rabbinic Judaism and its limitation") without giving up on ethnocentrism, national election, and purity of identity. As Elizabeth Sauer observes, "Milton is no less interested than Cromwell in preserving the purity of English identity and the native tongue, in establishing the role of the English as the elect, and in providing England with a national myth by developing a seamless continuity between Hebrew and Christian history."[72]

But, despite its successful conflation and subordination of the Hebraic/Judaic to the Christian, Milton's vision of the redeemed nation as mandated by providential history remains troubled by unresolved tensions between the two traditions, especially their opposing vantage points on the body. These tensions can be felt in the subtle ambivalences that unsettle the ultimately progressive vision of Christian history set out in Books 11 and 12. Against the current of the last books' compelling narrative drive toward the post-dynastic, post-apocalyptic maternal future, Adam sustains his carnal, literal-mindedness and his emphasis on the body and corporeal genealogy until the very end. As we have seen, he clings to these even after he begins to grasp Michael's Protestant typological allegory.[73] Adam's reluctance to embrace typological allegory fully may reflect something of Milton's own strong reluctance to give up entirely on the paradisal

[72] Elizabeth Sauer, "Religious Toleration and Imperial Intolerance," *Milton and the Imperial Vision*, p. 223.

[73] See Thomas Luxon, *Literal Figures: Puritan Allegory and the Reformation Crisis in Representation* (Chicago: University of Chicago Press, 1995), and Jeffrey S. Shoulson, *Milton and the Rabbis*.

interchange between spirit and body, thought and action, and the monism that informs the middle books of his epic when God and Nature bid the same: when freely chosen virtue did not abstract the reasoning human subject from natural time and space, but instead allowed for the refinement of social, intellectual, and sexual pleasure in material, this-world, terms, "Improved by tract of time" (5: 498).

My special aim in this chapter has been to demonstrate that Milton's ambivalence about the body and the Hebraic is the same ambivalence that is foundational to *Paradise Lost*'s early modern conception of the reformed nation. For Milton, the reformed nation is spiritually all-inclusive, but it is also an elect and exclusive collectivity. It is a Judeo-Christian polity, but one in which Judaic particularity, exclusiveness, and embodiment at once foreshadow the special promise of the reformed Christian nation and are cancelled by the spiritual fulfillment of that promise. More generally, and perhaps more importantly, Milton's epic illuminates a hitherto largely obscured dimension of the concept of the modern nation and its signature Janus-face: that this national idea equally depends upon and represses Judaism's insistent emphasis on the body, separation, historical presence, and memory.

As already noted, Eve nourishes the conceptual paradoxes that organize the new nation in *Paradise Lost* because her relationship to the new era and the new Israel is far less fraught than Adam's, as her concluding speech demonstrates. Her joyful final speech smoothly reverses her brief but poignant lament in Book 11 when she overhears Michael's injunction that Adam and Eve must leave Eden. Just as the thought of death drives Adam to lament the futility of producing progeny after the Fall, the prospect of eternal exile compels Eve to question the need to "leave/Thee native soil" and the nursery of flowers that she "bred up with tender hand/ From the first op'ning bud, and gave ye names." "Who now," she asks, "shall rear ye to the Sun, or rank/Your tribes, and water from th'ambrosial fount?" (11: 270, 276–79). Yet another example of the maternal "tender"- ness topos that, as noted in Chapter 3, is a defining feature of the guide- books and their reformist vision of the English family/nation, Eve's lament recapitulates the concerns about nurture and surrogacy, maternal breast- feeding and wet-nursing that make the maternal breast a fiercely conten- tious early modern cultural site, over which the battles between bloodline and merit, biological and affective measures of attachment and inherit- ance, class and national identity were fought.

Michael does not answer Eve's questions about who will nurse her brood of flowers—who will supply them with "water from the ambrosial fount." Instead, he gently reminds her: "Lament not.... with thee goes/ Thy husband, him to follow thou art bound;/Where he abides, think there thy native soil" (11: 287–92). Eve has no further lines until her final speech, after Michael's and Adam's descent from the "hill/Of Paradise the highest," (11: 377–8) when she reveals that the dream vision she receives from God amplifies her understanding of Michael's equation between her native soil and her husband. Echoing her love song in Book 5, the newly awakened Eve declares her perfect devotion to Adam in a circular expression of love—from "thou" to "thou": "thou to mee/Art all things under Heav'n, all places thou" (12: 617–18). Rosenblatt reads Eve's final speech as proclaiming "the continuity of her love for Adam and the supreme value of their life together" and argues that "her message is entirely antipathetic to Paul's."[74] Yet, while Eve re-dedicates herself to Adam by reiterating her circular expression of love from Book 4, she also distances herself from Adam when she proclaims her future role as the maternal nurturer of the new era. Although Eve begins her speech as the exemplary Hebrew wife, or the *Uxor Ebraica* of John Selden's text,[75] she ends her speech as the exemplary Christian mother.[76] In her lament in Book 11, Eve worries about who will nurse her growing brood of fruits and flowers. In her final speech, she happily shares her good news about the redemptive future. Eve's gospel or, by the commonplace seventeenth-century metaphor, her new spiritual milk replaces the ambrosial water with which she had nourished her natural fruits and flowers in Eden. The Apostles will serve up the same nourishing gospel message when they set out to evangelize the nations.

As the exemplary convert, Adam hears Eve's evangelical message with pleasure, but, as we note once again, he remains silent after he hears her good news. His silence conveys the subtle distance that opens up in the end between Adam and Eve or the old biological father and the new spiritual mother. As his natural wife, Eve cleaves resolutely to Adam; but, as "second Eve," she is also spiritually cleft from him, breaking the natural bond to which Adam mistakenly clings after the Fall. This is why Adam and Eve's departure from Paradise, and *Paradise Lost*, is paradoxical: both "hand-in-hand" (12: 648) and "solitary" (12: 649).[77] In political terms, the spiritual, maternal arc of the revised future divorces restoration from human patrimony, dynastic succession, the laws of primogeniture, and

---

[74] Rosenblatt, *Torah and Law in Paradise Lost*, p. 59.

[75] John Selden, *Uxor Ebraica, seu De Nuptiis & Divortiis ex Iure Civili, Id Est, Divino & Talmudico, Veterum Ebraeorum* (London: Richard Bishop, 1646).

[76] Rosenblatt, *Torah and Law in Paradise Lost*, p. 59; on the great influence of *Uxor Ebraica* on Milton, pp. 87–90. See also Rosenblatt's magisterial *Renaissance England's Chief Rabbi: John Selden* (Oxford: Oxford University Press, 2006), Chapters 3, 4, 6.

[77] Sauer persuasively interprets these lines rather differently: "the final image in the poem is of the shared experience of Adam and Eve wandering together "in sorrow forth." "Milton and some female contemporaries," *Milton and Gender*, ed. Catherine Gimelli Martin (Cambridge: Cambridge University Press, 2004), p. 148.

the paternal line. In the end, true restoration will be achieved by those disenfranchised by the restored Stuart court: the unsung heroes and right-eous political and Protestant martyrs who patiently await the redeemed future, despite their terrible suffering in the actual present. With Adam's final silence, the fallen natural father yields to the ascendant spiritual mother; the relentlessly tragic drama of history finds its comic, Christian conclusion. Adam's silence reinforces his revolutionary "female" insight: "by small/Accomplishing great things, by things deem'd weak/Subverting worldly strong" (12: 566–68).

On the one hand, then, through Michael's instruction, Adam learns to internalize the new nourishing m(other). But, on the other hand, his final silence, which is gently imposed by the archangel, "for now too nigh/ Th' Archangel stood" (12: 625–6) also highlights the obscure fault lines between his lost embodied past and his happier spiritual future. Adam both does and does not traverse the boundary between the old and the new dispensations. Although he embraces his new, nourishing male-ma-ternal subjectivity and ratifies Eve's prophetic vision of restoration as spir-itual unity, he also remains, for Milton, the obsolete father of the old organic order—the patriarch of the dynastic nation.

It is precisely this patriarchal Adam that is celebrated by such royalist political treatises as Robert Filmer's *Patriarchia*. For Filmer, Adam's God-given command over creation justifies the absolute sovereignty and the divine-right of kings:

> This lordship which Adam by command had over the whole world, and by right descending from the patriarchs did enjoy, was as large and ample as the absolutist domain of any monarch which hath been since the creation.[78]

In 1652 Filmer had challenged Milton by name in his *Observations con-cerning the originall of government upon M. Hobs Leviathan, M. Milton against Salmasius, H. Grotius, De jure belli, Mr. Hunton's Treatise of Mon-archy*. In his epic, Milton pushes against and subdues the Adam of Filmer and other royalists, but without directly impugning our first father's per-fection before the Fall. Perhaps the references to "small," "weak," and "simply meek," in Adam's final response to Michael in Book 12 might be read as diminishing Filmer's "large and ample" Adam. In the same vein, Adam's feminization in Milton's lines may also undercut the absolute, patriarchal power justified by Filmer's royalist Adam. For these reasons, among others, Milton denies Adam the privilege and authority of having the last word in his epic.

---

[78]   Thomas Filmer, *Patriarcha: Or the Natural Power of Kings* (London, 1680), p. 13.

Thus, although the reformed Adam silently says yes to the new order and to the death of the old dynastic nation, his silent affirmation also shares space with the silence that suppresses but does not fully cancel out the unregenerate Adam. This is the Hebraic Adam of Books 11 and 12 who is stubbornly attached to the body, Judaic literalism, and Hebrew monism—and who resists Pauline foreclosure on material history, sexuality, physical progeny, particularity, and concrete social practices and customs. Adam's final silence keeps both the old Hebraic Adam and the new Christian one in ambiguous play, in much the same way that, as we have seen, Milton's prescient vision of the modern nation's enabling contradictions simultaneously recognizes and suppresses the body, the particular, and the genealogical.

As we shall see next, in *Samson Agonistes*, Milton reiterates and revises the ambivalences that unsettle his epic and its modern vision of the reformed Christian nation. In *Paradise Lost*, the new Eve nurtures the modern nation's foundational contradictions by mediating between the exclusive and the inclusive, the body and the spirit, and the Hebraic and the Christian. In *Samson*, however, the hero must reject both Dalila's and Manoa's proffers of maternal nurture in order to make the new nation coherent or conceivable in Judeo-Christian terms. Samson not only must release Israel/England from Philistine bondage (idolatry and tyranny) from without, but he also must break his nation's bondage to the law, the body, and the old body politic from within. In *Samson*, Milton takes a new international approach to Judeo-Christian history and to the role that maternal nurture plays in the renovation and sustenance of godly community.

5

"I was his nursling once":
Internationalism and "nurture holy"
in *Samson Agonistes*

At the end of his long lament on his "miserable change" from exalted Hebrew champion to debased Philistine slave, Samson remembers: "I was his nursling once" (*SA*, 634). Samson's recollection poignantly highlights the deity's nurturing, maternal aspects. We are reminded of the great "Nourisher" (*PL*, 5: 398) of the prelapsarian universe in *Paradise Lost* and the terrible spiritual deprivation that Adam and Eve experience after the Fall. Seduced into betraying his "capital secret" (394) and sold into bondage by Dalila, the fallen hero feels cut off from God's special nurture, much as a child who is too abruptly weaned from his mother's breast. Samson's depiction of himself as a nursling denied God's milk should be familiar now from Puritan sermon literature, in which penitent sinners are routinely compared to "a sucking Infant depending on the Breasts of Divine Providence."[1] As we also have seen, the same discourse of nursing innovatively and conservatively expresses the associations between maternal nurture and national identity, and the private and public realms. Just as Samson's self-image as the deity's rejected nursling encapsulates the darkness that floods the blind hero's soul ("O dark, dark, dark, amid the blaze of noon, /Irrecoverably dark, total eclipse/Without all hope of day!" [80–2]), it also carries a conspicuous but curiously under-examined political charge. I argue that the image of Samson as God's nursling and the allusions to nursing and nurture elsewhere in the drama open a new window onto Milton's preoccupations in *Samson Agonistes* with national reformation and the reformed national subject.

As in *Paradise Lost*, in *Samson Agonistes*, Milton views the interplay between nation and nurture through the lens of Judeo-Christian scriptural history. His perceptions of the maternal-centered household and its relationship to male, public affairs and the nation closely intertwine with his fraught understanding of Hebraic precedent, Judaic law, and

[1] Samuel Willard, *A Compleat Body of Divinity* (Boston, 1726), pp. 32, 131.

Hebrew–Christian affinities and disparities. Surprisingly, although Milton's Hebraism in *Samson* has been well documented,[2] little attention has been paid either to the maternal matters implied by Samson's self-definition as God's nursling or their relationship to *Samson's* Hebraism and its vision of Israel/England's reformation.[3] This omission is particularly notable when we consider the close connections between Samson's "nurture holy" (362) and his prophesied role as Israel's emancipator. For Samson to maintain his Nazarite election and thereby fulfill his destiny as his nation's champion, he must drink only "the clear milky juice" (550) sanctioned by God. As Manoa observes, "For this did the angel twice descend? For this/Ordained thy nurture holy" (361–2) and "caused a fountain at thy prayer/From the dry ground to spring, thy thirst to allay/After the brunt of battle" (581–3). As these lines suggest, Samson's identity as Israel's champion is rooted in his "nurture holy" (362), the "clear milky juice" (551), and the sacred fount offered by a nursing deity. Samson's God is a nursing father—a figure discussed in Chapter 3. By highlighting the deity's maternal nurture, Milton evokes the celebrated trope of the nursing father from Isaiah 49:23 and Numbers 11:12, as well as the medieval image of Jesus as a nurse, as in the writings of Julian of Norwich, already noted. But, Milton also suggestively departs from his biblical source in Judges by dissociating Samson's "nurture holy" (362) from Samson's mother. In *Samson*, the poet suppresses the figure of Samson's mother, despite the central role that she plays in her son's special upbringing in the Judges narrative. Thus, although Milton takes pains to present Samson's God as a nurturing mother, he makes an equally motivated effort to keep Samson's human mother out of his text.[4] As a result, a divide between sacred and human forms of maternal nurture opens up in the text—a gap that profoundly shapes *Samson's* characterization of the hero as Israel's "great deliverer" (279).

---

[2] Achsah Guibbory, *Ceremony and Community from Herbert to Milton*, Chapter 7; Michael Lieb, "'Our Living Dread': The God of *Samson Agonistes*," *Milton Studies* 33 (1997): 3–25, and "'A Thousand Fore-Skins': Circumcision, Violence, and Selfhood in Milton," *Milton Studies* 38 (2000): 198–219; Jason P. Rosenblatt, "Samson's Sacrifice," in *Form and Reform in Renaissance England: Essays in Honor of Barbara Kiefer Lewalski*, ed. Amy Boesky and Mary Crane (Newark: University of Delaware, 2001), pp. 321–37; and Jeffrey S. Shoulson, "Epilogue: Toward Interpreting the Hebraism of *Samson Agonistes*," *Milton and the Rabbis: Hebraism, Hellenism, and Christianity*, pp. 240–62.

[3] A notable exception is Dayton Haskin's excellent "The Father's House (and the Suppression of Samson's Mother)," in *Milton and the Burden of Interpretation* (Philadelphia: University of Pennsylvania Press, 1994), pp. 138–46.

[4] Samson's mother/Manoah's wife is not named in Judges, but she was known to the Babylonian rabbis as "Zlelponi" or "Zlelponith." See Tamar Kadori, "Wife of Manoach: Samson's Mother: Midrash and Aggadah," *Jewish Women: A Comprehensive Historical Encyclopedia* (Online), http://jwa...org.../wife of manoach-samsons mother-midrash and aggadah.

Milton's occlusion of Samson's mother and the maternal body distances *Samson Agonistes* from the wide range of texts—from Richard Brathwaite's *The English Gentleman and English Gentlewoman* to Spenser's *A View*—which assign nursing mothers and the maternal-centered domestic sphere a crucial role in shaping national identity. As noted in Chapter 3, the discourse on maternal breast-feeding situated in Puritan didactic literature on household government in texts such as Gouge's *Of Domesticall Duties* has special relevance in this context, since the attention that Gouge and other Puritan domestic guidebook authors pay to the nurture of babies accords exactly with Milton's own understanding in *Of Education* that the nurture of children "from the cradle" is "worth many considerations." At the end of his tract, Milton states that the primacy of nurture is a subject that he himself would have addressed "if brevity had not been my scope" (*CPW*, 2: 414–5).

While *Of Education* places Milton in sympathetic dialogue with writers such as Gouge, *Samson Agonistes* shows that the mature Milton's assessment of the social and national value of maternal nurture differs considerably from the position adopted in the guidebooks. As we saw in Chapter 1, the guidebooks celebrate the identity-forming power of maternal nurture modeled in Hebraic scriptural ideals of nursing motherhood, such as Sarah and Jochobed. By contrast, Milton's *Samson* not only occludes Hebraic models like Samson's mother, but also the idea that purity of self is constituted by pure nurture—the same monist idea of pure selfhood celebrated by both the Puritan guidebooks and their Hebraic paradigms of nursing motherhood. Samson undermines the Puritan/Hebraic equation between pure maternal nurture and pure identity when he drinks from Dalila's "fair enchanted cup" (934) rather than from God's sacred fount and the "clear milky juice" (550) sanctioned by the Law. Placing Dalila over the deity, the bad breast over the good one, Samson evacuates the ideal of selfhood that he had hoped to embody by adhering to the Nazarites' strict code and Judaic law.

At the beginning of the drama, Samson bitterly laments his lapse; he compares himself to "a foolish pilot" who has "shipwrecked/My vessel trusted to me from above/Gloriously rigged: and for a word, a tear, /Fool' have divulged the secret gift of God/To a deceitful woman" (198–201). Dalila is, by contrast, "Like the stately ship.../With all her bravery on, and tackle trim" (714–17). Although Samson could fell "a thousand foreskins" in Ramath-lehi (144–5), he is easily deceived by Dalila and Israel, both of whom sell him into Philistine bondage. In the end, however, the poet asks us to see Samson's "foolish pilot[ing]" (198) as inaugurating both the hero's and his nation's emancipation. When Samson drinks from Dalila's "enchanted cup" (934), he eradicates "what once I was" (22).

Samson's transgression obliterates his former identity, but it also initiates his new one. Although Samson wishes to "pay on my punishment; /And expiate, if possible, my crime, /Shameful garrulity" (489–91), he does not know at first how to do so. He angrily laments his debasement ("to grind in brazen fetters" [35]) and decries his blindness. Nevertheless, he accepts "the will/Of highest dispensation, which herein/Haply had ends above my reach to know" (60–2). As we shall see, these mysterious "ends" (62) are revealed, appropriately enough, "in the close" (1748). In the meantime, however, Samson senses that Manoa's and Dalila's offers to ransom and nurse him will not expiate him. "Spare that proposal, father, save the trouble/Of that solicitation" (487–8), he tells Manoa.

To console the despairing hero, the Chorus tells Samson that, despite his transgression, God has exempted him "From national obstriction, without taint/Of sin, or legal debt" (312–13). True to the perfect contrariness of highest justice (about which I shall have more to say soon), lawbreaking can sometimes accord with the Law. At first Samson fails to understand how breaking the Law may also be a way of obeying the Law. Later, however, at the pivotal moment of insight that he inwardly experiences during his encounter with the Philistine Officer, he recognizes that, in special instances, when, for example, Israel's liberty is at stake, the Law legitimates transgression. At first, Samson refuses to break the Second Commandment against worshipping idols and says "no" three times to the Philistine decree that he perform at the Dagonalia. But, then, *volte face*, he says "yes" three times. Despite the commandment against worshipping idols, he agrees to participate in the Philistines' celebration of Dagon. He also claims that he will "Nothing to do, be sure, that may dishonor/Our law, or stain my vow of Nazarite" (1385–6). "Yet this be sure," he repeats, "in nothing to comply/Scandalous or forbidden in our Law" (1408–9). A third time, he says, "of me expect to hear/Nothing dishonorable, impure, unworthy/Our God, our law, my nation, or myself" (1423–5). This third reply also is Samson's last directly spoken dialogue in the drama.

That Samson might not have to "pay on [his] punishment" (489) or that his crimes might be exempt from "legal debt" (313) should not completely surprise us.[5] By marrying Samson, Dalila has been granted a related legal exemption under the Hebraic code. When she becomes

---

[5] For an opposing view, see John Guillory, "The Father's House: *Samson Agonistes* in its Historical Moment," in *Re-membering Milton: Essays on the Texts and Traditions*, ed. Mary Nyquist and Margaret W. Ferguson (New York and London: Methuen, 1987), p. 165. Guillory argues that Samson's dispensation from "the *constituting prohibitions* of Hebraic culture" makes his conflicts about obeying or transgressing God's law by participating in the Dagonalia seem "trivial."

Samson's wife, she is acquitted of the charge that she is unclean—in apparent violation of the prohibition against intermarriage articulated in Deut. 7:3, "You shall not intermarry with them [Hittites, Girgashites, Amorites, Canaanites, and so forth]: do not give your daughters to their sons or take their daughters for your sons." Rather than Deuteronomy, Milton in *Samson* invokes another, seemingly contradictory scriptural account of intermarriage. He alludes to Moses's two marriages to foreign wives: the first one to the Midian woman, Zipporah (Exodus 2), and the second one to the Cushite woman (Numbers 12:1). The case of Zipporah is rather ambiguous. The Priestly genealogy that begins Genesis 25:1–4 indicates that the Midians are related to the Hebrews through Abraham's second wife, Keturah. Midian is the fourth son of Abraham and Keturah; thus the Midianite Zipporah may not be an outsider after all. The case of the Cushite woman, however, is different—with significant implications for Milton's *Samson*. According to Josephus, Moses promises to marry Tharbis, the daughter of the king of the Ethiopians (or Cushites) if she agrees to deliver up her city to his army. "No sooner was the agreement made, but it took effect immediately. Moses...gave thanks to God and consummated his marriage."[6] Both Miriam and Aaron criticize Moses for marrying the Cushite woman, but in the end they acknowledge that the marriage between their father and the Ethiopian princess is sanctioned by God. Unlike Deut. 7.3, the story of Moses and the Cushite woman/ Ethiopian princess in Numbers legitimates intermarriage. It implies that, at moments of national crisis, the ancient Hebrews accepted intermarriage, albeit with conditions: the foreign spouse must uphold Mosaic Law and in no way interfere with the Hebrew spouse's worship of God. Samson alludes to these conditions, when he asserts that should a man and woman of different nationalities marry, the wife must give up her own nationality and instead adopt her husband's national identity: "Being a wife, for me thou wast to leave/Parents and country." As we shall see soon, these lines also assert the primacy of natural law.

Like the story of Moses, *Samson* comments on the special legality of intermarriage and other prohibitions at an especially tense moment in the history of the biblical Israel: during the Philistine occupation. Although the Judges narrative is unclear about whether or not Dalila is Samson's wife or his concubine, Milton's drama explicitly identifies Samson and Dalila as a married couple. Just as Miriam and Aaron at first cannot accept their father's foreign wife, so Manoa at first cannot approve his son's marriages to Philistine women. He has nothing to say about Samson's claim that "Divine impulsion" (422) prompted him to marry Philistine women:

---

[6] Josephus, *Antiquities of the Jews*, p. 58.

"I state not that" (424). Additionally, neither he nor the Chorus comments on Samson's argument that his first marriage serves as the legal precedent for his second one: "I thought it was lawful from my former act, /And the same end" (231–2). Not unlike the Chorus, however, Manoa does believe that God will forgive Samson for marrying Dalila and exempt him from legal debt: "God will relent, and quit thee all his debt" (509). Despite appearances, the deity still supports his champion: "His might continues in thee not for naught" (588). The similarity between Manoa's vision of Samson's ultimate redemption and the Exodus and Numbers stories of Moses's emancipation of the Hebrews from the Egyptians and other oppressors is striking. Just as Moses marries a foreign woman (or, possibly, two) and leads his people to the Promised Land, so Samson, despite (or because of) his foreign marriages, is sanctioned to deliver Israel from Philistine occupation. One notable difference, however, is that, unlike Moses's Cushite wife, Dalila fails to recognize that by marrying a Hebrew she has automatically placed herself under the Law, with all its stipulations. Not only does she refuse to give up her loyalties to her native country, but she also seduces her spouse into breaking his religious vows.

Dalila's betrayal of Samson as his Hebrew wife—and *not* her uncleanness as a foreigner—provides the hero with legal grounds for divorce: "Thou and I long since are twain" (929). This almost completely overlooked detail—that Dalila's foreignness per se does not negate her legitimacy as Samson's Hebrew wife—is highly significant. In addition to underscoring the complexities of the Mosaic Law, Milton highlights the legality of Samson's international marriage to Dalila to argue covertly against the Crown that adherence to the Law of Nations is in the national interest. Milton explicitly makes this same argument in *Of True Religion*.[7] Published almost contemporaneously with *Samson Agonistes*, Milton's last prose tract asserts that national reformation ultimately must yield to "the conscience of Nations": the principle of international law, rights, and liberties. In 1668 the Cavalier Parliament voted to enforce the Clarendon Code's harsh penal laws against dissenters and Nonconformists. In 1670 the king formed an alliance with Catholic France against Holland, resulting in the Third Anglo-Dutch War. On June 1, Charles and Louis XIV (who were first cousins) signed the secret Treaty of Dover, which stipulated that France would bail out England financially in exchange for

---

[7] On Milton's nationalism and *Of True Religion*, see Elizabeth Sauer, "Milton's *Of True Religion*, Protestant Nationhood, and the Negotiation of Liberty," *Milton Quarterly* 40.1 (2006): 1–19, and Paul Stevens, "How Milton's Nationalism Works: Globalization and the Possibilities of Positive Nationalism," *Early Modern Nationalism and Milton's Nation* (Toronto: University of Toronto Press, 2008), pp. 273–304.

English troops and Charles's promise to convert to Roman Catholicism. In 1672 Charles II issued the second Declaration of Indulgence, in which he offered to repeal or reduce penal sentences for dissenters and Roman Catholics. Milton and like-minded compatriots viewed the king's new policy on toleration as a politically calculated first step toward officially re-establishing Roman Catholicism in England.

Published in 1672, *Of True Religion* responds to Parliament's stepped-up repression of Nonconformists, Charles II's toleration of Catholics and dissenters, and the king's alliance with France. Rather than extend liberties to Catholics, Milton argues that toleration should be granted comprehensively to all forms of English Protestantism, from the most conservative to the most radical. A united Protestant front would make England less susceptible to the real-and-present danger of Catholic influence. The papacy, he argues, holds "the conscience of Nations" in complete disregard. Milton observes that "the Pope [...] pretends right to kingdoms and states, and especially to this of *England*, Thrones and Un-thrones Kings, and absolves the people from their obedience to law, sometimes interdicts to whole Nations and the public worship of God, shutting up their Churches" (*CPW*, 8: 429). He thus argues for broader toleration of English Protestants at the same time that he strongly agitates against religious freedom for English Catholics. To underscore England's difference from Roman Catholic nations, however, the poet also urges Parliament to uphold international law, under which foreign-born Catholics residing in England must be granted religious freedom: "Forreigners [are] Privileg'd by the Law of Nations" (*CPW*, 8: 431). The Law of Nations grants exemptions from civic law in special cases—not unlike God's higher law in *Samson Agonistes*. Milton's pamphlet had no specific impact on public policy. Nevertheless, when Parliament reconvened in late 1672, it denounced the king's Declaration and refused to fund the ongoing Third Anglo-Dutch War. Charles II was forced to comply.

Samson's national and international vision relies on a set of interlocking ideas (civil rights and the Law of Nations, natives and foreigners) very similar to those articulated in *Of True Religion*. Not unlike *Of True Religion*, *Samson* reflects Milton's disappointment with nationalist and partisan politics, especially his disaffection with the English people, after the regicide, for giving up the very same liberties for which they had formerly clamored.[8] After the Restoration, Milton's "*Patria*" crosses national borders: "One's *Patria* is wherever it is well with him" (*CPW*, 8: 4). *Samson* reflects this same

---

[8]　See Victoria Kahn, "Disappointed Nationalism: Milton in the Context of Seventeenth-Century Debates about the Nation-State," in *Early Modern Nationalism and Milton's England*, ed. David Loewenstein and Paul Stevens (Toronto: University of Toronto Press, 2008), pp. 249–72.

international sentiment, as does the poet's association of Samson's and Dalila's marriage with "the law of nature, law of nations (890)."[9]

Milton's conception of the Law of Nations owes much to the legal writings of Hugo Grotius.[10] Grotius's consideration of international law
emerges in response to the Thirty Years War, with its seemingly ceaseless
conflicts between Catholic and Protestant nations. In 1625, distressed by
Europe's continual religious-and-national antagonisms, Grotius argues in
De Jure Belli ac Pacis for the primacy of international law. (Milton met the
famous Dutch jurist in Paris in the mid-1630s, when the poet visited
the city during his Continental tour; he also explicitly praises Grotius at
the beginning of the *Doctrine and Discipline of Divorce*.)[11] Grotius maintains that, whatever their local differences, all Christian nations ultimately
are united by universal principles of justice, based on natural law and
right reason, such as the right to self-defense. These principles are para-
mount: "the law of nations is a more extensive right [than a civil right or
the authority of parents over children, masters over servants], deriving its
authority from the consent of all, or at least many nations."[12]

As Jason Rosenblatt documents, John Selden makes an equally significant contribution to the period's theorizing of the law of nature and the
law of nations in *De jure naturali et gentium justa disciplinam Ebraeorum*
(which Milton praises in *Areopagtica*). Selden bases his vision of international law on the seven universal commandments of the covenant that
God made with Noah and which were observed by Noah's gentile children (Noachides). To minimize the differences between the gentile
Noachides and the Jewish descendents of Moses, born under the law,
Selden emphasizes that the Hebrew Bible considers converts to Judaism
as identical to those born into the religion. When "a convert as well as a
freedman...comes into Judaism...they say that he was regarded as a 're-
cently born baby,' that he became Jewish and thus the things which were
altogether in the past, such as blood and affinity, were poured away."[13]

---

[9] For the Restoration context of *Samson*, see Sharon Achinstein, *Literature and Dissent in Milton's England* (Cambridge: Cambridge University Press, 2003), pp. 133–53.

[10] See Elizabeth Oldman, "Milton, Grotius, and the Law of War: A Reading of 'Paradise Regained' and 'Samson Agonistes'," *SP*, 104.3 (Summer 2007): 340–75. Julie Stone Peters discusses Grotius's "just war" theory in relationship to *Paradise Lost* in "A 'Bridge Over Chaos': *De Juri Belli*, *Paradise Lost*, Terror, Sovereignty, Globalism, and the Modern Law of Nations," *Comparative Literature* (Fall 2005): 373–93.

[11] Barbara K. Lewalski, *The Life of John Milton* (Oxford: Blackwell, 2003), p. 89. Milton describes Grotius as "a most learned man...whom I ardently desired to meet" (*CPW*, 4.1, 615); cited in Lewalski, p. 89.

[12] Hugo Grotius, *The Rights of War and Peace, including the Law Of Nature and of Nations*, trans. Richard Tuck (Ithaca: Cornell University Press, 2009; The Liberty Fund), p. 25.

[13] John Selden, *Uxor Ebraica* (1646), 2.18, p. 207; quoted in Rosenblatt, *Renaissance England's Chief Rabbi: John Selden* (Oxford: Oxford University Press, 2006), p. 102.

Milton draws specifically on Selden's ideas about the relations between gentiles and Jews, foreigners and natives, and conversion and naturalization under the Mosaic Code in *Samson Agonistes*. As already noted, Milton's Samson accuses Dalila of deliberately forgetting that when she became a Hebrew by marriage, she was required to give up her loyalty to her native country: "if aught against my life/Thy country sought of thee, it sought unjustly, /Against the law of nature, law of nations, /No more thy country" (888–91). As Rosenblatt astutely points out, Samson's denunciation of Dalila alludes to the title of Selden's *De juri naturali et gentium* ("the law of nature and nations").[14] As we shall see soon, not unlike Dalila, Samson also must break all ties to his native country before he can fulfill his prophesied role as Israel's redeemer.

For different reasons, the Law of Nations also was successfully invoked in England in 1608 by Thomas Ellesmere in *Calvin's Case* or *The Case of the Post-nati*. Ellesmere argues that King James's Scottish subjects, who were born after the royal union of 1603, were equally subjects of England and could own property there. Like Grotius, Ellesmere considers the Law of Nations to be universal, pertaining exclusively to, in Ellesmere's terms, "the realms of Christian kingdoms and empires."[15] In *Samson*, Milton also links international law with Christian universalism. As just mentioned, however, his idea of "the law of nature, law of nations" (890) additionally is mediated by specific civil-war conflations of the Law of Nations and right reason. As Barbara Lewalski observes, "The Delilah episode asks readers to reaffirm the principle that justified the Revolution—that natural law takes precedence over civil and ecclesiastical authority."[16] Not unlike Grotius, Solicitor General John Cook, the legal mastermind behind the regicide, maintains in *King Charls, his case* that the "general law of all nations" is inseparable from "the unanimous consent of all rational men in the world."[17] Cook's formulation of the Law of Nations, however, was expressly designed to strengthen the regicides' case against the king. The new High Court of Justice ruled that right reason and the Law of Nations overrode loyalty oaths sworn to Charles I. Under the Law of Nations, such oaths pertained only to popish attempts against Charles's kingship. As interpreted by Cook, the Law of Nations ruled that loyalty oaths could not prevent Parliament and the English people either from

[14] I am indebted to Jason Rosenblatt, *Renaissance England's Chief Rabbi*, esp. Chapters 4 and 6.

[15] Louis A. Knafla, *Law and Politics in Jacobean England: The Tracts of Lord Chancellor Ellesmere* (Cambridge: Cambridge University Press, 1977).

[16] Lewalski, p. 531.

[17] John Cook, *King Charls his case: or, An appeal to all rational men, concerning his trial at the High Court of Justice, 1649* (Thomason/84: E.542[3]), p. 22. See Geoffrey Robertson, *The Tyrannicide Brief* (Chatto & Windus, London, 2005), esp. p. 173.

accusing the king of treason or proceeding with capital punishment should the king be found guilty as charged. This ruling cleared the way for Charles's trial and execution, for which no legal precedent could be found in English common law.

Milton compresses a considerable amount of classical and modern jurisprudence in *Samson Agonistes*—far too much adequately to account for in this chapter. I nevertheless wish to make the following two-part claim. First, in *Samson*, Milton, like Grotius, Selden, and Cook, identifies godly community with all right-reasoning men in the world; and, second, the poet's international vision closely coheres in as-yet unrecognized ways with *Samson's* repudiation of the private household (the mother's house) as impossibly co-opted. Once an alternative domain where the work of Reformation could proceed in secret, the maternal household, drained of its reforming energies, now collaborated with the state—as Milton experienced first-hand in the years immediately following the Restoration.[18] Rather than uphold the Hebraic/Puritan maternal model for generating national subjects, Milton dramatizes the emergence of a new, un-co-opted, international space of selfhood, where true Reformation could proceed in spirit, uniting godly, rational men the world over.

In *An Apology... Against Smectymnuus*, Milton writes that "I conceav'd my selfe to be now not as mine own person but as a member incorporate into that truth whereof I was perswaded."[19] Milton's distinction between the self as "mine own person" and "as member incorporate" can help us to refine our understanding of the alternative, international personhood that Samson must acquire in order to fulfill his divine prophesy.[20] To experience the transformation from "mine own person" to a "member incorporate into that truth," Samson must give up his native attachments to the old nation of Israel. As just noted, Dalila is required, but fails, to make a similar kind of sacrifice when she marries Samson. By offering to nurse him, however, both Dalila and Manoa tempt the hero to return to the embodied particularity of the old family/nation. To resist the monist-maternal definition of purity, Samson must "other" and "mother" himself in the image of the nurturing deity. The hero becomes, not unlike Milton, a foreigner in his native land, but, at least potentially, he (again like the poet) re-experiences his self as "member incorporate" of a higher, more exclusive, international union. Samson exchanges his old national self for his new international one by accomplishing what Adam could

---

[18] Lewalski, *The Life of John Milton*, p. 403.

[19] John Milton, "An Apology...against Smectymnuus," *The Works of John Milton*, 3.1.284, gen. ed. Frank Allen Patterson (New York: Columbia University, 1931).

[20] See Joanna Piciotto, "The Public Person and the Play of Fact," *Representations* (Feb 2009) 105.1: 86.

not: he breaks the "link of nature" (*PL*, 9: 914) that binds him by blood, law, and genealogy to his father, Dalila, and the biblical Israel. Samson's self-sacrifice is even more impressive when we consider how flexible the Law can sometimes prove to be. As already noted, the Law can even accommodate transgression. By successfully resisting the Law, even (or especially) in its most charitable, accommodating form, Samson is poised for inclusion in the exclusive community of "all rational men in the world."

Of equal importance for my purposes here is that, as just noted, Milton's foreclosure on the private, maternal-centered home in *Samson* represents a crucial, although hitherto overlooked, manifestation of the poet's post-Restoration turn toward internationalism. Rather than a private refuge from tyranny, the private, maternal-centered household (Dalila's house and Manoa's house) in *Samson Agonistes* replicates not only Philistine oppression but also bondage to the co-opted Israel and Mosaic Law— even in its most charitable form. As Samson admits, despite his heroic feats as God's champion on the battlefield, he was a slave to his wife and his own inner weakness. To serve his maternal deity, Samson must reject both of his natural, nursing "mothers," Dalila and Manoa—in sharp contrast to Milton's Jesus in *Paradise Regained*, whose mission to free humankind from the bonds of sin is closely tied to the "mother's house private" (*PR*, 4: 639). Appearing side by side in the 1671 volume, *Paradise Regain'd... To which is added Samson Agonistes*, *Paradise Regained* and *Samson Agonistes* force attention to the differences and similarities between the maternal realm in the new and old dispensations. A peripheral site in terms of plot, Mary's private household forms the spiritual center of *Paradise Regained*: it is the origin of and endpoint to Jesus's "unrecorded" (*PR*, 1: 16), "unsung" (*PR*, 1: 17), heroic acts of resistance in the wilderness to satanic temptation. Whereas in *Samson*, the private world of domesticity thwarts the hero's spiritual emancipation and re-dedication to God, in *Paradise Regained*, the mother's home opens up new, emancipating personal and social spaces: maternal alternatives to governing patriarchal institutions. In contrast to the corrupt public realm, with its repressive priests and tyrannical kings, the "mother's house private" (*PR*, 4: 639), because "deem'd... obscure" (*PR*, 1: 23–4), allows the work of national liberation to continue in secret by planting the spiritual seeds of resistance to oppression in the regenerate heart and by nurturing their growth. Unlike the redemptive private maternal household in *Paradise Regained*, the nurturing home in *Samson Agonistes* threatens Samson's salvific mission. To fulfill his providentially inspired role as Israel's emancipator, he must free himself from the old family/nation of "carnal Israel"; more specifically, he must repudiate the Hebraic/Puritan equation between maternal nurture and pure identity, and acquire a new international understanding of true selfhood and community.[21]

As we shall see, for Milton, resistance to the old natural bonds that organize the biblical Israel requires an almost superhuman demonstration of inner strength. At first, the deeply conflicted, melancholic Samson lacks the spiritual fortitude and personal integrity to negate the monist link between pure nurture and national identity, and to cut the powerful natural and maternal ties that closely bind him to Dalila, Manoa, and Israel. Unable to give up on his earthly nursing mothers and mother-country, Samson is not ready to re-dedicate himself to his divine nursing Mother–Father or to incorporate himself spiritually into the universal community of true believers. Samson in his opening soliloquy exists in a painful state of suspended animation. He can no longer remain the same, but he is not ready to change: he is too late to live within the law, but too early to be redeemed by grace.

Milton thus not only highlights the difference between gospel and law by publishing *Paradise Regained* and *Samson Agonistes* in one volume, but he also illuminates the painful liminal space *between* gospel and law, the new and old dispensations.[22] Samson miserably shuttles back and forth between his legendary but outmoded heroic past and his providentially inspired but uncertain future. Some respite from his profoundly unsettling, to-and-fro, state of mind comes to the hero when he rejects both Dalila's and Manoa's proffers of domestic nursing. With this rejection, he is able for the first time to dissociate pure identity from pure maternal nurture, thus undermining the Hebraic/Puritan model for generating natural, national subjects. By refusing nursing from Dalila and Manoa, Samson not only breaks his kinship bonds and his blood ties to the old Israel, but he also begins to break free of the painful irresolution that imprisons his spirit. Samson must not only emancipate himself from Philistine rule but also from Hebrew law and its nursing, maternal

---

[21] In contrasting "the carnal Israel" or the Israel "according to the flesh" with the new spiritual Israel in Christ, Augustine in *Tractatus adversus Judaos* quotes Paul's statement in 1 Corinthians 10:18, which itself is a comment on the allusion to Israel's apostasy in Deut. 32.17: "Behold Israel according to the flesh."

[22] Joseph Wittreich argues that Milton's Samson represents a negative heroic example, to be contrasted with Jesus in *Paradise Regained*; see *Interpreting Samson Agonistes* (Princeton: Princeton University Press, 1986) and *Shifting Contexts: Reinterpreting Samson Agonistes* (Pittsburgh: Duquesne University Press, 2002). Guibbory underscores the complimentariness between Milton's indictments of Restoration England and his representations of Jesus and Samson as prophets and emancipators in *Paradise Regained* and *Samson Agonistes* in *Christian Identity, Jews, and Israel*, pp. 275–91. For Loewenstein, Milton positively portrays Samson as a "radical saint" who, like oppressed, post-Restoration dissenters, exercises his liberty of conscience despite his terrible adversities; see *Representing Revolution in Milton and his Contemporaries: Religion, Politics, and Polemics in Radical Puritanism* (Cambridge: Cambridge UP, 2001), pp. 269–91.

paradigm for ensuring national purity. Rather than separate, opposing peoples, one idolatrous, the other godly, the Philistines and the Hebrews are identical: both thwart the hero's emancipation. This indifference to national differences coheres with the emphasis on intermarriage and internationalism in *Samson*. Whereas Milton in his earlier works had identified England as the new Israel, in *Samson Agonistes*, he associates God's nation with the international community of the Protestant elect.

But if Milton rejects the pure monist self associated with Puritan/ Hebrew models of the nursing mother, he also refuses to resuscitate the hyper-masculine selfhood associated with the traditional, action-driven hero.[23] This route is occluded by the close associations in his writings between patriarchy and the abuses of monarchist, papist, and other corrupt institutional forms of male power in his prose works. Samson's heroic feats on the battlefield do not emancipate him from his inner servitude. In political terms, Samson's dual rejection of Dalila and Manoa doubly forecloses on Restoration renewals of traditional patriarchal forms of monarchical government and Reformist attempts to make natural motherhood a central measure of non-dynastic nationhood. In *Samson*, all partisan or particularized measures of nationhood yield in the end to a higher worldwide collectivity, to which all regenerate individuals, regardless of their local, ethnic, or national differences, could uniformly pledge their spiritual allegiance. It is this godly international union and not the chosen nation of Israel/England that Samson ultimately serves in Milton's drama.

In addition to Grotius's and Selden's commentaries on the Law of Nations, Milton's godly internationalism reflects the new global consciousness that provides seventeenth-century English culture with much of its reorganizing and reformist energies. As James Holstun and Bruce McLeod point out, Milton traveled in intellectual circles and was, indeed, at the center of a literary and political culture that (partly in response to England's anxieties about the rapidly changing nature of its cultural and territorial borders), sought new means for reconciling every aspect of experience into one unifying global system, along the lines of Bacon's *Novum Organum*.[24] Throughout the seventeenth century, as Latin began to lose its position as the sole written European language of scholarship, linguistic reformers, including Bacon, Jan Amos Comenius, John Webster, and John Wilkins, among others, proposed various schemes for a universal language that would restore the linguistic unity that mankind had enjoyed

---

[23] See Mary Beth Rose, *Gender and Heroism in Early Modern Literature*, Chapter 4.

[24] Bruce McLeod, "The 'Lordly Eye': Milton and the Strategic Geography of Empire," in *Milton and the Imperial Vision*, pp. 48–66, and James Holstun, *A Rational Millennium: Puritan Utopias of Seventeenth-Century England and America* (New York: Oxford University Press, 1987), esp. pp. 110–15.

before Babel.[25] Like the language projectors, early modern cartographers, such as Willem Janzaoon Blaeu and Jodocus Hondious, and travel writers, such as Richard Hakluyt and Samuel Purchas, endeavored to place all near-and-far regions and peoples of the world on the same map with England or Europe at the center. Purchas sees himself as providing "a Prospective Glasse, by which thou maist take easie and neere view of those remote Regions, Peoples, Rites, and Religions."[26] The Hartlib circle similarly imagined England at the vanguard of "Universal Reform" in its proposals for the recolonizing of Ireland.[27] Utopists like James Harrington maintained that universal Reformation could be achieved by a systematic rationalization of English culture and society:

> the balance of a commonwealth that is equal is of such nature that whatever falleth into her empire must fall equally, and if the whole earth fall into your scales it must fall equally; and so you may be a greater people and yet not swerve from your principles one hair.[28]

A correlative desire for equitable reform the world over shapes the staging of Israel's deliverance in *Samson*. Milton links Samson's delivering of Israel from Philistine bondage to the struggle for freedom from ungodly forms of power in every sphere of experience—in the private maternal-centered household as well as on the masculine, public stage.[29] In *Areopagitica*, Milton allegorizes the newborn English republic as an arisen female Samson, shaking her locks and staring with undazzled eyes into the sun. In *Samson*, the allegory of the new Israel as a phoenix-like Samson forms the spiritual climax of Milton's drama of national-and-selfreformation, one that unfolds both outside and inside of the hero. The inner recesses of the mind, the private household, and the public stage: all are arenas in which Samson must free himself and his nation from ungodly rule. In *Samson*, Milton, not unlike Samuel Purchas, provides a "Prospective Glasse" capable of uniting all aspects of Reformation, inward and

[25] See David S. Katz, "Babel Revers'd: The Search for the Universal Language and the Glorification of Hebrew," *Philo-Semitism and the Readmission of the Jew to England, 1603–55* (Oxford: Clarendon Press, 1982), Chapter 2, and Sharon Achinstein, "The Politics of Babel in the English Revolution," in *Pamphlet Wars: Prose in the English Revolution*, ed. James Holstun (London: Frank Cass, 1992), pp. 14–44.

[26] Samuel Purchas, *Purchas His Pilgrims, or Hakluytus Posthumus*, 20 vols. (Glasgow: James Maclehose and Sons, 1940), 4.420.

[27] Cited in McLeod, p. 51.

[28] James Harrington, *The Political Works of James Harrington*, ed. J.G.A. Pocock (Cambridge: Cambridge University Press, 1977), p. 322. For an illuminating analysis of Harrington's utopianism, see James Holstun, *A Rational Millennium: Puritan Utopias of Seventeenth-Century England and America* (New York and Oxford: Oxford University Press, 1987), Chapter 4.

[29] See *Ceremony and Community*, p. 186,

outward, into a single global vision. Samson struggles for independence from both the nurturing maternal-centered home and the male world of public affairs—against the Philistine Other and within his Hebrew self. Rather than the new Israel/England, Milton looks forward to an international community of regenerate spirits.

Milton's political vision of a new international "Israel" finds theological justification in Paul's transcendent vision of Christian community in Galatians 3:28: "There is neither Jew nor Greek, there is neither bond nor free, there is neither male nor female, for ye are all one in Christ Jesus." This prescription negates the binary oppositions between embodied measures of identity (ethnicity, class, and gender) and resolves them into a higher spiritual community: the all-ness of unity in Christ. Pauline Christianity provides a legitimating theological framework for the Protestant vision of worldwide reform that shapes Milton's drama. In Pauline fashion, *Samson* proclaims that the deliverance of Israel/England from ungodly Philistine rule is indissolubly linked to nothing less than the spiritual Reformation of *every* kind of pre-existing sphere of experience. Since all is fallen, all must be redeemed. This Pauline model of English identity as at once universal and Protestant lies behind not just Milton's strange conflation of Dalila and Manoa as maternal nurturers, but also his allusions to nursing and his revaluation of the Hebrew Bible's maternal means of ensuring pure identity in monist terms.

What makes *Samson*'s apparent disparities (such as the opposition between the Hebrews and the Philistines) contiguous is that Pauline Christianity depends upon an enabling contradiction between inclusion and exclusion or the same "exclusive universalism,"[30] which, as already noted, underpins Grotius's and Ellesmere's conceptions of the Law of Nations. For Paul, the Christian church is equally open to all (it is indifferent to gender, ethnicity, and class), but it also represses those who refuse to participate in its all-expansiveness. As Boyarin observes,

> If the Pauline move has within it the possibility of breaking out of the tribal allegiances and commitments to one's own family, it also contains seeds of an imperialist and colonizing missionary practice. The very emphasis on a universalism, expressed as concern for all of the families of the world, turns rapidly (if not necessarily) into a doctrine that they must all become part of our family of the spirit, with all of the horrifying practices against Jews and other Others that Christian Europe produced.[31]

[30] My debt here is to Paul Stevens's brilliant analysis of "exclusive universalism" in " 'Leviticus Thinking' and the Rhetoric of Early Modern Colonialism," *Criticism* 35 (1993): 441.

[31] Daniel Boyarin, *Carnal Israel: Reading Sex in Talmudic Culture*, p. 234.

This same exclusive–inclusive Pauline imperative finds expression not only in Milton's suppression of both Samson's mother and the Hebrew Bible's embodied, maternal-nursing measures of national identity, but also in his attempt to transmute received notions of nation, nurture, and the Hebraic into a reformed vision of universalism, inspired by Galatians 3:28 and upheld by the Law of Nations. Milton's model of godly internationalism reflects a tolerationist vision of global union. At the same time, this tolerant universal order cannot tolerate those forms of identity, such as the Hebraic, which resist Milton's Pauline mandate of inclusiveness by insisting upon remaining spiritually and corporeally separate from Christendom. By breaking the "bond of nature" (*PL*, 9: 956) when he rejects Dalila's and Manoa's proffers of maternal nurture, Samson sacrifices his ethnic, tribal, and kinship allegiances to the old Israel. For Milton, Samson's sacrifice exemplifies the one that all godly persons must make to find inclusion within reformed Christendom's universal family/nation of the spirit.

Odd contiguities between inclusion and exclusion, renewal and repression, the nation and the international thus frame Milton's overlapping preoccupations with maternal nurture, the Hebraic, and the Judaic, in *Samson Agonistes*. Milton's allusions to nursing and nurture are enmeshed in his drama's matrix of seemingly disparate concerns, including the spiritualization of motherhood, the ambivalent status of the biblical Israel, and the poet's Pauline vision of international Protestantism.[32] Taking the drama's Protestant universalism as its overarching frame, I next trace the complex, comprehensive vision of Reformation that shapes Milton's drama by examining first in sequence and then together the interlocking Hebraic and maternal implications of the drama's preoccupations with nation and nurture.

HIGHEST JUSTICE AND GOD'S CONTRADICTING LAW

Despite their seeming disparities, Samson's identity as God's nursling and his prophesied role as Israel's "great deliverer" (279) are intimately connected facets of *Samson*'s radical, Pauline vision of universal reformation—a vision that profoundly affects the play's rendering of its Hebraic sources and the Judaic past. Samson's "nurture holy" (362) provides a central focus for Milton's efforts to assimilate the Judaic to the Christian:

---

[32] On the interconnectedness of Milton's nationalism and internationalism, see Linda Gergerson, "Colonials Write the Nation," p. 169.

to make Judaic law serve and serviceable to Christian liberty. At the start of the drama, Samson's prophesied role as national redeemer depends upon his strict adherence as a Nazarite to the deity's dietary prohibitions against wine and all unclean foods. In accordance with the Nazarite code and its maternal measures of pure identity, Samson must fulfill his destiny as Israel's "great deliverer" (279) by drinking only the "clear milky juice" (551) that anoints him as God's nursling and grants him his incredible physical strength. As we have seen, Samson's prophesied purification of Israel's national body is at first inseparably linked to Samson's own clear, milky purity as a Nazarite. By the end, however, to deliver himself and Israel from Philistine bondage and to return to the deity's breast, the hero must reject not only his nursing mothers, Dalila and Manoa, but also Hebrew scripture's nursing paradigm of pure identity: the monist linkage between pure nurture and pure selfhood that is prescribed by the Nazarite code and the Law. As already noted, Samson must break the Law in order newly to legitimate his own integrity. God's perfect contrariness sanctions these strange identities among lawbreaking, purity, personal integrity, and highest justice. Samson must reject his nursing mothers (Dalila, Manoa, and the charitable Mosaic Code) to renew his special status as God's nursling. As the Chorus states, only those "who doubt his ways not just" find God's edits "contradicting" (300, 301).

For the Chorus, contradiction is compatible with divine justice, as in the rabbinic view. According to the famous statement from the Babylonian Talmud, a heavenly oracle settled the disputes between the opposing Houses of Hillel and Shammai by declaring that "these and these are the words of the Living God": "R. Abba said Shmuel said: 'The House of Hillel and the House of Shammai disputed for three years.' These said, 'The *halakah* is according to us,' and those said, 'The *halakah* is according to us.' A heavenly voice went out and said, 'These and these are the words of the living God.'"[33] That Milton adopts a charitable nurturing rabbinic view of Judaic law is borne out when the Chorus emphasizes that the "contradicting" (301) nature of God's "own edicts" are "Just.../And justifiable to men" (293–4)—lines that echo *Paradise Lost*'s central objective: to "justify the ways of God to men" (*PL*, 1: 26). In *Samson*, as almost everywhere else in his writings, Milton reaches back to an earlier Christian notion of contradiction, dissent, and debate as compatible with unity

---

[33] See Jason P. Rosenblatt, "Samson's Sacrifice," and Jeffrey S. Shoulson, "Epilogue: Towards an Interpretation of *Samson Agonistes*." Quoted in Daniel Boyarin, "'These and These are the Words of the Living God': Apophatic Rabbinism," paper, Dartmouth Regional Seminar in Jewish Studies, Hanover, NH, May 2002.

and justice. Richard Lim observes that the orthodox Christian emphasis on *homonoia* or simplicity was a late development.[34] Unlike Augustine, Origen celebrates sectarianism as compatible with Christian community and church government. Origen's position on the compatibility between debate and consensus, contradiction and justice, is almost identical to that espoused by early Palestinian rabbinism. In response to political and social shifts in late-Roman Christian society, however, the Nicene bishops came to embrace simplicity rather than debate as the church's central doctrine, and endorsed "the notion that there is and always had been only one truth and the social ideal is *homonoia*, total agreement without discussion or dispute." To preserve its integrity, rabbinic Judaism "went in the opposite direction from orthodox Christianity": it not only "rejected *homonoia* but promulgated instead a sensibility of the ultimate contingency of all truth claims."[35]

Through the Chorus, Milton ratifies as consistent with reformed Christianity the rabbinic view of the compatible and, finally, incomprehensible relations between contradiction and justice, heterogeneity and social order. This is why Samson can claim that his marriage to the woman of Timna "motioned was of God" (222), and why he can conclude (erroneously) that his second marriage to Dalila is "lawful from my former act" (231). "There seem to be occasions proscribed by the law," as Shoulson points out, "when such cultural and ethnic cross-pollinations have a salvific potential: 'for by occasion hence/I might begin *Israel*'s Deliverance'" (224–5).[36] Milton asks us to treat the Chorus's charitable equation between justice and contradiction, community and difference, with utmost seriousness rather than dismiss their view as an expression of their intellectual and ethical limitations. We can hear the Milton of the divorce tracts in the Chorus's denunciation of those who strictly identify God with the Law. Rather than associate God with the free expression of Truth, they would force divine justice into received forms: they "confine th'interminable, /And tie him to his own prescript, /Who made our laws to bind us, not himself" (307–9). These carnal formalists fail to recognize that God is the Law, but he cannot be confined by the Law: "For with his own laws he can best dispense" (314).

Yet if it greatly extends our understanding of the relations among justice, contradiction, and the flexibility of the Law, in the end the Chorus nevertheless only offers partial insight into why God's "contradicting"

---

[34] Richard Lim, *Public Disputation: Power and Social Order in Late Antiquity* (Berkeley and Los Angeles: University of California Press, 1994).

[35] Daniel Boyarin, "'These and These are the Words of the Living God': Apophatic Rabbinism," pp. 3–5.

[36] Shoulson, pp. 224–5.

(301) enforcement and suspension of the Law is the highest form of justice. The Chorus's rabbinic view of the charitable contradictions within the Law and divine justice is conjunctive: "these *and* these are the words of the living God." Rather than promulgate "a sensibility of the ultimate contingency of all truth claims," by celebrating contradiction as compatible with justice, as do the rabbis,[37] Milton understands contradiction as a measure of higher union, as does Paul in Galatians 3:28: "for ye are all one in Christ Jesus." Unlike his Chorus, Milton echoes Paul when he implies that divine justice goes beyond good and evil, transcending, rather than conjoining, these categorical oppositions. Rather than provide a charitable and expansive, both/and formulation of divine justice, as does rabbinic Judaism, Milton offers a transforming, neither/nor vision, indebted to his category-bursting, sublime-and-Pauline ideal of community as "neither Jew nor Greek." The lines in *Samson* between the Pauline neither/nor and the rabbinic both/and vision of divine justice and reformed community can seem impossibly confused. Samson's heroic task is not so much to distinguish the transforming, Pauline perspective of divine justice and community from the expansive rabbinic view, although he must do this too. His task is to demonstrate that a neither/nor Pauline principle of justice and community overlaps with but also supersedes the rabbinic monist understanding of the contradictory nature of God's edicts that is espoused by the Chorus.

These Judeo-Christian concordances and disparities find their clearest expression in relation to *Samson*'s concerns with the "nurture holy" (362). By imbibing only "clear milky juice" (551), Samson conforms to Hebrew scripture's monist ideal of pure identity, as put into practice by Sarah's and Jocobed's breast-feeding of their elect sons. But, when Samson breaks the prohibitions against wine and unclean food by drinking from Dalila's "enchanted cup" (934), he falls away from Hebrew scripture's monism and into the painful dualism that he decries in his opening soliloquy: the terrible disjunction between his "Immeasureable strength" (206) and his "wisdom nothing" (207). His "servile toil" (5) at the "common prison" (6) at Gaza brings "Ease to the body some, none to the mind" (18). "O impotence of mind, in body strong!" he laments (52). Most painful to Samson is the disparity he perceives between "what once I was" (22), the perfect embodiment of the Hebraic/Puritan monist equation between pure nurture and pure self, and "what am now," an unhinged body and mind (22).

In their reminiscences of Samson's heroic past, the Chorus and Manoa echo Samson's despair at his broken self and life: "O change beyond

[37]   Boyarin, "Apophatic Rabbinism," pp. 3–5.

report, thought, or belief!/...Can this be he, /That heroic, that re-
nowned/Irresistible Samson?" laments the Chorus (117, 124–5). "O
miserable change!" cries Manoa. "Is this the man,/That invincible
Samson?" (340–1). Like Samson in his soliloquy, the Chorus underscores the monist equation between pure nurture and pure identity:
Samson's resistance to the "Desire of wine and all delicious drinks"
(541) marks his difference from "many a famous warrior" (542). "[N]or
did," the Chorus adds, "the dancing ruby/Sparkling, out-poured, the
flavor, or the smell, /Or taste that cheers the heart of gods and men, /
Allure thee from the cool crystalline stream" (543–6). With its gem-like
"ruby/Sparkling," wine falsely "cheers" the vulgar "heart" (545) and
"fills with fumes" (552) idolatrous heads. The Semichorus contrasts the
maternal equation between pure drink and pure identity with drunken
acts of Philistine idolatry at the end of the drama when they observe
that the Philistine elite at the Dagonalia: "Drunk with idolatry, drunk
with wine (1670)...Unwittingly importuned/Their own destruction to
come speedy upon them" (1680–1). "O madness," proclaims the
Chorus, "to think use of strongest wines/And strongest drinks our chief
support of health/When God with these forbidd'n made choice to rear/
His mighty champion, strong above compare, /Whose drink was only
from the liquid brook" (553–8).

Despite their expansive understanding of the Law, however, neither
Manoa nor the Chorus can see beyond the Law;[38] both prove unable to
foresee how the gap that opens up between what Samson once was and
what he is after his transgression provides the new international subjectivity necessary for the hero's ultimate, inward, phoenix-like rebirth. In
the end, Manoa acknowledges neither a gap in nor a transformation of
Samson's heroic identity: "Samson hath quit himself/Like Samson"
(1709–10). Like the lamenting Adam in Book 10 of *Paradise Lost*, however, Samson is caught for most of the drama in the liminal time-space
between the two dispensations, between law and grace, and between his
natural role as his nurturing father's biological son and his spiritual role as
the chosen son of the nursing deity. Unable to understand how his lapse
prepares him for his spiritual regeneration, the fallen hero, like Manoa
and the Chorus, rails against his failure to sustain a Hebrew-monist unity
of body and mind: "But what availed this temperance [against unclean
drink], not complete/Against another object more enticing?" (558–9).
Manoa echoes his son's failure to comprehend how God's ordaining of

---

[38] Stanley Fish reads Manoa as limited by his desire to rationalize the indeterminacies in
Samson's story in "Spectacle and Evidence in *Samson Agonistes*," *Critical Inquiry* 15 (1989):
556–86.

Samson's special nurture is related to his prophesied role as Israel's deliverer. "For this," Manoa wonders, "did the angel twice descend? For this/ Ordained thy nurture holy" (361–2)? Was it not "God who caused a fountain at thy prayer/From the dry ground to spring, thy thirst to allay" (581–2)? Manoa is bewildered by what he perceives as the terribly unjust equation between crime and punishment that appears to result from God's "contradicting" (301) justice: "Alas methinks whom God hath chosen once /To worthiest deed, if he through frailty err, /He should not so o'erwhelm, and as a thrall /Subject him to foul indignities" (368–71). Like Manoa, the Chorus, despite their earlier embrace of contradiction as compatible with divine justice, later questions the seemingly inequitable ways in which God "Temper'st thy providence through [man's] short course, /Not evenly" (670–1), such that "Just and unjust, alike seem miserable" (703).

Although at first Milton seems to incline more towards the Chorus's rabbinic affirmation of contradiction as divine justice than to its skeptical rejection of God's "hand so … /contrarious" (668–9), in the end, he associates the limits of the charitable rabbinic view with the Chorus's skepticism. Whether Samson upholds the prohibitions against impure drink and unclean women or breaks them (i.e. whether he is the clean Samson of the past or the unclean Samson of the present), his purity and that of Israel are guaranteed, since God's dispensing of the laws and his dispensing with them are equally holy acts, producing the same result: purity. Ultimately, for Milton, however, the purity achieved by abrogating Law transcends the purity gained by adherence to the Law. Purity in *Samson* is at first tightly tethered to but finally unmoored from the monist equation between pure nurture and pure self, even when that equation is given an expansive, charitable rabbinic interpretation, accommodating taint, difference, and transgression. Pure identity transcends the Law even in its most expansive and flexible form; it is a measure of the higher formlessness, or spirit, of the Law, which the Son sets free when he does away with genealogy, circumcision, dietary prohibitions, and other "carnal" Judaic measures of pure identity. As noted earlier, even the Law itself provides exemptions from its own rulings. Ultimately, Milton dissociates highest justice from Judaic /Hebraic notions of justice, even at their most charitable and expansive, even as he also demonstrates the very close proximity between the Pauline and the rabbinic perspectives on divine justice. While Milton never completely gives up on Mosaic Law, he nevertheless suggests that a Pauline abrogation of or dispensing with the Law produces the higher, spiritual forms of justice and purity that he believes should govern the new international-Protestant "Israel," which is the Israel that Samson in spirit ultimately serves.

NURSING SAMSON: OCCLUDING
THE NATURAL MOTHER

Samson's preoccupations with the hero's "nurture holy" (362) thus refract Milton's desire to synthesize and subsume the rabbis' charitable interpretation of Judaic law within the Pauline doctrine of love. The play's domestic ideology betrays the same unifying and universalizing impulses. By linking the hero's "nurture holy" (362) to his nation's election, his imbibing of "clear milky juice" (551) to the imprinting of his own and Israel's identity, *Samson* prompts comparison with many of the texts, such as William Gouge's *Of Domesticall Duties*, that we have examined in earlier chapters, in which nursing and nurturing motherhood assume distinctly national dimensions. Although the materials on nursing discussed in earlier chapters date mostly to the early 1600s (although many were reprinted frequently over the course of the seventeenth century), interest in the national import of maternal nursing and nurture did not diminish in any way after the Restoration. English cultural preoccupations with nurture, nursing, and nationhood find much more explicit and institutional expression after the Restoration and well into the eighteenth century, when maternal breast-feeding is bureaucratized.[39] In underscoring the connections between nurture and nation, Milton's drama intervenes in the period's ongoing polemicizing of the domestic sphere and of maternal nurture, more specifically.

As already noted, in *Of Education*, Milton is in sympathetic dialogue with writers such as Gouge, but his reluctance to write about the primacy of nurture also underscores his distance from them. That distance becomes wider and much more explicit in *Samson*—a point that is intimately intertwined with Milton's ambivalence about the particularity of Judaic law, and his emphasis on Protestant universality in his drama. In the guidebooks, as discussed in Chapter 1, Gouge, Cleaver, and Dod, among other Puritan writers on household government, celebrate the identity-forming power of maternal nurture by alluding to Hebraic scriptural models of nursing motherhood, specifically, Sarah and Jochobed. Like the guidebook writers, Milton in *Samson* is clearly interested in the relations between nurture and identity, but, unlike them, he takes pains not to represent these relations as legitimated by Hebraic scriptural models of maternal nurture.

Samson's natural mother, who plays a significant role in Samson's upbringing in the Judges narrative, is nowhere to be found in Milton's

---

[39] Ruth Perry, "Colonizing the Breast: Sexuality and Maternity in Eighteenth-Century England": 195–200.

drama. In Judges, the angelic visitor first appears to Manoah's wife alone to tell her that, if she abstains from wine and all unclean foods, she will bear a child. The angel commands Samson's mother never to cut her child's hair; she must rear Samson as a Nazarite and in this way begin the delivery of Israel from the Philistines. On a second visit, the angel repeats these commands to both Manoah and to his wife. These plot details are written out of Milton's tragedy, such that neither Samson's mother's special distinction as the first to be visited by the angel and to hear his commandment nor her divinely mandated abstinence, nor her role in the special nurturing of Samson is acknowledged in Milton's tragedy. While in Judges, Samson's mother is charged with her son's Nazarite upbringing, in *Samson*, God alone "rear[s] /His mighty Champion." Milton's Manoa states that the angel did "twice descend" (361) but he makes no distinction between the angel's first visit to his wife alone and the second visit to them both; indeed, he states that he alone—"I"—"prayed for children" (352) and "I gained a son" (353). The Chorus also alludes to the angel's visitations, when, in bidding farewell to Samson after he agrees to perform at the Dagonalia, they pray that the angel that announced his birth will reappear and "stand /Fast by thy side" (1431–2) when the hero enters the Philistine theater. Like Manoa, the Chorus notes "thy father's field," (1432) but omits all references to Samson's mother:

> Send thee the angel of thy birth, to stand
> Fast by thy side, who from thy father's field
> Rode up in flames after his message told
> Of thy conception, and be now a shield
> Of fire.
>
> (1431–5)

Providing yet another example of this occlusion, Samson never speaks directly of his mother. At most he obliquely refers to "my parents" (25) or to a gender-neutral "the womb" (634). Instead of a metaphor of maternal fruitfulness, as in the case of the prelapsarian Eve, the womb is a metonym for Samson's inborn, providential destiny: "I was his nursling once and choice delight, /His destined from the womb" (633–4).

The occlusion of Samson's mother in Milton's retelling of the Samson story de-emphasizes the monist equation between pure milk and pure identity celebrated by the guidebooks and other texts. Seductively replicating official forms of domination, the maternal-centered household in *Samson* no longer exists as a viable alternative to ungodly government, as an obscure site where the reformation of the nation and national subject could be cultivated through maternal nurture, as imagined in the guidebooks and the other texts we have considered. Instead of the private household, *Samson* imagines a new international subjective space in which to stage the epochal, conceptual shift from the Hebraic paradigm of the chosen nation to the Pauline model of international Protestantism.[40]

Like Adam, who struggles to learn from Michael how to combine the ruined monism of the lost Paradise with the hopeless dualism of the new era, Samson at first experiences his split mind and body as the painful, punitive mark of his failure to adhere to the Nazarite code of nurture. In the end, however, he comes to reinterpret his shattering of the maternal, Hebraic/Puritan bond between pure nurture and pure self as leading to a higher form of reintegration that is motioned by God. If at first Samson experiences his mind–body split as a terrible wound, he later understands this duality as preliminary to the material rebirth of his spirit. No longer in thrall to fear, pity, desire, or other passions, the hero embraces the free over the formal way to truth, allegorized by his fiery, phoenix-like rebirth:

> But he though blind of sight
> Despised and thought extinguished quite,
> With inward eyes illuminated
> His fiery virtue roused
> From under ashes into sudden flame.
>   (1687–91)

Samson's inward deliverance by the "sudden flame" of his "fiery virtue" subjectively rehearses Israel's emancipation not only from Philistine bondage, but also from Judaic law and the nursing mothers and fathers of Hebrew scripture.

Milton's excision of Samson's mother along with Samson's dual rejection of Manoa and Dalila are thus the keys to the reformation of the hero's selfhood. Drinking "pure milky juice" (551) represents a crucial measure of Samson's special cleanness and election as a Nazarite. But Milton suggests that giving up this embodied maternal conception of "nurture holy" (362) allows Samson not only to "divorce" his wife and father, but also to "divorce" himself: to sever himself from "what once he was,"(22) a Hebrew self. As a phoenix, he re-begets his self and Israel by delivering his spirit and that of his nation from the tyranny of the body and the old body politic (the Philistine/Stuart nation). His purified spirit is poised for incorporation within the new exclusive-universal body politic: the spiritual body of true believers. Through Samson, Milton converts Hebraic monism into Christian terms. Through its occlusion and repudiation of the maternal-centered household, the drama allows us to glimpse the new Israel as a

---

[40] Raymond Williams, *The Year 2000* (New York: Pantheon, 1983).

worldwide corporation of emancipated Protestant selves. Milton's Pauline universalism negates the embeddedness of identity in the embodied particularities of nation, class, and gender and reidentifies the reformed self as capable of mixing completely with regenerate selves the world over. Raphael's description of how angels make love offers a vivid depiction of Milton's Christian-materialist sense of the spiritual self as "member incorporate": Because no "obstacle... /Of membrane, joint or limb, exclusive bars: / Easier than air with air if spirits embrace, /Total they mix, union of pure with pure /Desiring" (*PL*, 8: 624–8).

NEITHER DALILA NOR MANOA: "HOME TO THY COUNTRY AND HIS SACRED HOUSE"

David Leverenz has established that domestic guidebook literature on maternal breast-feeding was largely penned by Puritans and intended for a mostly reformist readership.[41] Given Puritan skepticism about birthright and bloodline as customary measures of class status and dynastic authority, it is not difficult, as we saw in earlier chapters, to understand the appeal that, especially by mid-century, the guidebooks' emphasis on mothers' milk as a natural and divinely mandated determinant of character must have had for reformers like Milton, eager to upend existing political hierarchies and root out the Roman magic and mystery in which official church-and-state power was shrouded. But, as we recall, the same shaping power over identity that Puritan writers increasingly accorded to mothers' milk helped to transform the maternal breast into a central object of not only the reformist political imagination, but of the royalist imaginary as well. The assumption that breast-milk was white blood helped to intensify royalist efforts to claim control of the nursing breast, as demonstrated by both *Basilikon Doron* and *Eikon Basilike*.

These battles for ownership of the maternal breast as a symbol of male political power and national identity are worth returning to in this context, since they offer an important, new framework within which to read Milton's ambiguous national figurations of the nurturing household. In *Samson*, the maternal breast is a complex but compromised symbolic site, as the conflation of Dalila and Manoa as maternal nurses makes clear. Rather than point up the marked differences between Philistia and Israel, the parallels between Dalila's and Manoa's offers to nurse Samson blur the distinctions between the subjugated Hebrews and their

[41] David Leverenz, *The Language of Puritan Feeling: An Exploration in Literature, Psychology, and Social History* (New Brunswick, NJ: Rutgers University Press), 1980, pp. 72–4.

Philistine overlords, underscoring the sameness rather than the difference of the two nations. In Hebrew scripture and post-biblical rabbinic commentary, circumcision represents *the* defining mark of Judaic particularity.[42] In rejecting both Dalila's and Manoa's attempts to nurse him, Samson underscores the likeness between Israel and her enemies and hence his need to assert his difference and independence not only from the foreskinned race but from the unforeskinned "Abraham's race" (29) as well.

But, while Samson must reject both Dalila and Manoa, Dalila's proffer of nurture appears at first more threatening to Samson than his father's, more difficult for the hero to resist, since he has already fallen prey to Dalila's seductive allure. But, in the end by resisting Dalila's seductive promise of "leisure and domestic ease" (917), Samson acquires the strength necessary to accomplish what Adam could not. In rejecting Dalila's proffer of nurture, he also breaks the "bond of nature" (*PL*, 9: 956).

Dalila's entrance follows in a highly motivated manner directly upon the Chorus's aforementioned skeptical discussion of the intemperate, uneven-handed justice that God meets out to his people, especially to those like Samson "thou has solemnly elected" (678). "Nor only dost degrade them, or remit /To life obscured, which were a fair dismission, /But throw'st them lower than thou didst exalt them high" (687–9). Toward these "thou oft," the Chorus observes, "Amidst their highth of noon / Changest thy countenance and thy hand with no regard /Of highest favors past" (682–5). As we have seen, during Manoa's visit with Samson, the Chorus celebrates the charitable expansiveness of the Law. But, as if anticipating the turning point provided by Dalila's visit, the Chorus shifts its focus from God's charitable justice to the horrific suffering that the deity's once-glorious champions are forced to endure—punishments that exceed the crime, that are "Too grievous for the trespass or omission" (691). If not butchered by the "hostile sword /Of heathen and profane" (692–3) and their carcasses left as prey for ravenous dogs and vultures, then they are fated to die a miserable death "in poverty /With sickness and disease thou bow'st them down" (697–8). Whereas the Chorus begins this speech by identifying patience as inward nurture, as "Some source of consolation from above; /Secret refreshings, that repair [man's] strength, / And fainting spirits uphold" (664–6), by the end, it points to the "evil end" (704) to which all men, patient and impatient, just and unjust, are miserably consigned. "Unseemly," the Chorus concludes, appears divine justice "in human eye" (690): "Just or unjust alike seem miserable, /For

---

[42] See Daniel Boyarin, *Carnal Israel*, pp. 7–8; Jeffrey S. Shoulson, *Milton and the Rabbis*, pp. 48–51.

oft alike, both come to evil end" (703–4). Entering after this litany of woes, Dalila provides the capstone to the Chorus's mounting list of grievances against a tyrannical God and to its extended repudiation of divine justice as intemperate and indifferent. Thus, in rejecting Dalila, Samson also resists the Chorus's skepticism about God's justice.

For the Chorus, God's justice is sinister rather than benevolent, since its false equity sentences "Just and unjust, alike" (703) to the same misery. The Chorus emphasizes the physical suffering and public humiliation that God's champions must endure. With Dalila's entrance, God's public degradation of man is extended into the domestic and private sphere. The "leisure and domestic ease" (917) that Dalila promises to confer upon Samson "At home" (917) privately replicates the public spectacle of misery to which all men are subjected. The rerouting of man's "evil end" (704) into the female-centered home is made through Dalila's offer to "intercede" (920) with the Philistine lords for Samson's release from bondage, relocating Samson from prison house to private house. Her desire to "intercede" (920) has priestly connotations, underscoring the idolatrous, Catholic duplicity implicit in her promise to attend to Samson with "*redoubled* love and care /With nursing diligence" (923–4, emphasis added). "[S]o bedecked, ornate, and gay" (712), she appears as a decorative double figure, made in the ornamental-oriental image of the Philistine "Sea-Idol," Dagon: "But who is this, what thing of sea or land?" (710). Dalila's house, like Dalila herself, blurs hitherto clear boundaries. While in *Paradise Regained*, "his mother's house private" (*PR*, 4: 639) represents a distinct alternative to the domination imposed by official rule, in *Samson*, this private maternal sphere not only replicates official rule, but also cunningly disguises it as a tender expression of love and nurture. Like the "double-mouthed" (971) fame that she does not "count it heinous to enjoy" (992), Dalila's expressions of amorous remorse would only "redoubl'[e]" (923) the public degradation that Samson now experiences. Rather than "Exempt [Samson] from many a care and chance to which /Eyesight exposes daily men abroad" (918–9), Dalila's "care /With nursing diligence" converts manly suffering into effeminate "domestic ease" (917, 924–5).

In offering to exempt Samson from the routine spectacle of misery to which divine justice seems to consign to all men, Dalila profanely echoes God's "full right to exempt" (310) Israel's champion from "National obstriction, without taint" (312). But whereas God's highest judgment equates exemption with emancipation, providing an antidote to "what remains past cure" (912), Dalila's desire to exempt Samson from public ridicule by nursing him at home would double the misery that, according to the skeptical Chorus, inevitably overwhelms God's champions. Rather than an alterative to the world of customary and official public affairs, the maternal-centered domestic sphere becomes yet one more arena that is co-opted by governing forms of despotic misrule. An ensnaring, but "stately ship /Of Tarsus" (714–5), Dalila is associated with Tarshiz, a symbol of pride in Psalm 48, and with the city on the River Cydnus where Cleopatra meets Antony. Suggestive of the "inborn disrespect for lawful, 'civilized' relations [that is] characteristic of Philistine women," as Mary Nyquist observes, Dalila, not unlike Shakespeare's Cleopatra, epitomizes the innately lawless and female nature of Oriental government.[43] Roman /Hebraic /republican discipline ("tackle trim" [717]) is covered over by the exotic, amber-scented perfume of Oriental or Egyptian /Philistine/Stuart decadence and effeminate slackness: "*Courted* by all the winds that hold them play, /And amber scent of odorous perfume /Her harbinger" (720–2, emphasis added).

Beyond its local application to the restored Stuart court, the "odorous perfume" (720) in the just-cited line also draws Dalila's "stately ship" (714) into the global arena by bringing into range the "spicy drugs" (*PL*, 2: 640) to which Milton, time and again, alludes in *Paradise Lost*, as when, for example, Satan in solitary flight from Hell is compared to the fleet of merchant ships sailing westward across the Indian Ocean:

> As when far off at sea a fleet descried
> Hangs in the clouds, by equinoctial winds
> Close sailing from Bengala, or the isles
> Of Ternate and Tidore, whence merchants bring
> Their spicy drugs; they on the trading flood
> Through the wide Ethiopian to the Cape
> Ply stemming nightly toward the pole. So seemed
> Far off the flying Fiend.
>
>   (*PL*, 2: 636–43)

Balachandra Rajan aptly observes that the "spicy drugs" (640) cited in this passage function as "a synecdoche for the entire range of conspicuous consumption" that Milton condemns.[44] Milton's figuration of Dalila as a trading ship places her among the merchant fleet with which Satan is associated by epic simile, inviting us to read the goods she carries, especially, her "nursing diligence" (924) in terms of the conspicuous consumption for which, as Rajan argues, "spicy drugs" (640) is a synecdoche. If as the "stately ship /Of Tarsus" (714–5) Dalila associates female, Philistine misrule with the orientalized decadence of an effeminate Stuart court, the same

---

[43] Mary Nyquist, "'Profuse, proud Cleopatra': 'Barbarism' and Female Rule in Early Modern English Republicanism," *Women's Studies* 24 (1994): 113.

[44] Balachandra Rajan, "Banyan Trees and Fig Leaves: Some Thoughts on Milton's India," in P.G. Stanwood, ed., *Of Poetry and Politics: New Essays on Milton and His World* (Binghamton, NY: Medieval and Renaissance Texts and Studies, 1995), p. 217.

metaphor also represents her as an adventuring merchant ship "With all her bravery on" (718) from Tarsus, the port on the Guadalquivir in Spain, "bound for th' isles /Of Javan or Gadire" (715–6), the Greek islands and the Spanish port of Cadiz. In addition to its anti-Spanish, anti-imperial nuances, this metaphor prompts us to insert the maternal gifts Dalila bears, "her nursing diligence" (924) and "enchanted cup" (934), within the framework of trade and its luxurious currency of spice, tea, exotic medicinals, and other Asian imports to England.

While a full discussion of *Samson*, the ideology of trade, and the currency of spice is quite clearly beyond the scope of this chapter, I nevertheless wish to comment on how Milton's figuration of the nursing Dalila as a merchant ship brings international trade and the maternal-centered household into surprisingly close proximity, despite their seemingly disparate spheres of influence. Equally important, Dalila's "ship" is a metonym for the commercial internationalism against which the godly internationalism espoused by *Samson* must push. In addition to maintaining domestic order, English gentlewomen played a central role both in "the absorption of the foreign necessitated by colonialism" and the formation of consuming subjects, hungry for foreign goods.[45] Milton suggests that, like sugar, spice, and tea, "nursing diligence" (924) encourages outlandish—not only foreign, but also luxurious and riotous—appetites from the breast, creating a primal, natural hunger for sugar and spices, such as those that perfume the wind powering Dalila's ship and the larger European merchant fleet, which Dalila's ship evokes. Despite its anticeremonial associations in the domestic guidebooks with nature and with social nakedness, nursing, in *Samson*, resembles such specifically female arts as the making of sugary confections or the ritualized performance of serving tea, through which the foreign was absorbed into the domestic sphere. In addition to its Circean implications, then, the association of Dalila's "nursing diligence" (924) with her "fair enchanted cup" (934) spotlights the female magic that enables this foreign–domestic exchange. Rather than purify and strengthen English identity, nursing in this context artfully translates foreign luxuries into domestic necessities. For Milton, the blurring of difference between luxury and necessity, foreignness and domesticity, within the "mother's house private" (*PR*, 4: 639) would impair the hero's struggle to emancipate himself: as already noted, Samson needs to estrange himself (to become foreign) from the old natu-

---

[45] Kim F. Hall, "Culinary Spaces, Colonial Spaces: The Gendering of Sugar in the Seventeenth Century," *Feminist Readings of Early Modern Culture: Emerging Subjects*, ed. Valerie Traub, M. Lindsay Kaplan, and Dympna Callaghan (Cambridge: Cambridge UP, 1996), p. 170. Elizabeth Kowaleski-Wallace, *Consuming Subjects: Women, Shopping, and Business in the Eighteenth Century* (New York: Columbia University Press, 1997).

ral family/nation order so that he can enter into the new spiritual/universal
one. Because one function of the domestic sphere is to translate the
foreign and alien into the domestic and native, Samson—in order to
preserve his spiritual estrangement from his nurturing mother-country—
must unequivocally reject Dalila's domesticating proffer of nursing.[46]

Manoa's proffer of nurture would similarly resuscitate Samson's natural, native Hebrew identity. Manoa believes that God will re-make Samson
into what he once was: the heroic champion of Israel. Rather than compel
him "to sit idle with so great a gift /Useless, and thence ridiculous about
him" (1500–1), Manoa would prepare Samson for his divinely appointed
moment of renewed active service to God: "I persuade me God had not
permitted /His strength again to grown up with his hair /Garrisoned
round him like a camp /Of faithful soldiery, were not his purpose /To use
him further in some great service" (1495–9). To which the Chorus responds: "Thy hopes are not ill founded nor seem vain, /Of his delivery,
and thy joy thereon /Conceived agreeable to a father's love" (1504–6).
But this exchange, with its hopeful linkage of Samson's restored divine
purpose to fatherly nurturance ("a Father's love" [1506]), is literally shat-
tered by the "hideous noise" (1509) made by Samson's rending of the
temple walls, the sound of "Blood, death, and deathful deeds" (1513), the
horrible death-cry of the crushed Philistines and of Samson himself. As
this highly ironic sequence of events dramatizes, Samson must bypass the
nursing domestic sphere, especially in its most utopian, paternal configu-
ration, in order to carry out his prophesied role as the providential agent
of Israel's deliverance.

Samson thus is doubly tempted by Dalila's and Manoa's nursing households to renew his heroic identity in outmoded terms. As a Hebrew
monist, Samson at first clings tightly to the identity between nature and
nation, inscribed by God's Law, but, in rejecting both Dalila's and Manoa's
offers to nurse him, Samson also implicitly acknowledges the obsolescence of the elect, natural bonds that organize the Hebrew family-nation.
He is thereby poised to become a "member incorporate" of the sublime-
international, neither-Jew-nor-Greek Israel that is reborn in Christ's resurrected body, as delineated in Galatians 3:28. Samson's estrangement
from his wife and father turns him into a foreigner—a stranger to his
Hebraic self and nation. Samson thus becomes a man (like the poet) without a natural identity and homeland. After collapsing the ethnic-national
differences between Hebrew and Philistine, however, the hero rediscovers
his holy "father's house" (1717) and returns home to it. Manoa looks

---

[46] Knoppers argues the early modern domestic sphere facilitates the exchange of luxury
into necessity and foreignness into Englishness in "Consuming Nations: Milton and
Luxury," *Early Modern Nationalism and Milton's England*, pp. 331–54.

forward to bringing Samson's body "Home to his father's house" (1734), where he will "build him /A monument" (1733–4) to which all "valiant youth" will make pilgrimage (1738). Manoa's idolatrous desire to turn Samson into a national icon makes his house unfit for the return of God's champion. Instead, Samson will return home to his divine nursing father. By underscoring this crucial distinction between Samson's nursing fathers, human and divine, Milton completely forecloses on the private household as a site of reformation.

The political valences of both Dalila's and Manoa's nursing households are unmistakable. Rather than open up an alternative site of reformed experience, the desire to nurse Samson instead replicates the master–slave relations that lock Israel and Philistia into a deadly embrace. Instead of releasing Samson from bondage, the nursing household would reduce the hero to what Milton in *The Tenure of Kings and Magistrates* terms a "slav[e] within doors," encouraging the inward slackness, "the inbred falsehood and wickedness," which make men love, rather than resist, the tyrannical kings and corrupt magistrates who abuse the power and authority "committed to them in trust from the people to the common good of them all." "Hence it is that tyrants are not oft offended…," Milton observes, "Consequently, neither do bad men hate tyrants, but have been always readiest with the falsified names of loyalty and obedience to colour over their base compliances" (*CPW*, 3: 237). Whereas in the domestic manuals and in Cromwell's speeches the rhetoric of maternal love and nurture serves to naturalize the artificial obligations imposed by a social-contract model of community, by contrast, "nursing diligence" (924) in *Samson* undermines this common league, draining men of the active reason needed to recognize international law, through which liberty, man's God-given birthright, although canceled by Adam's transgression, can be restored.

Samson's encounter with Harapha furthers the hero's estrangement from his Hebraic self and the biblical Israel. In Harapha's eyes, Samson has been reduced to a common "murderer, a revolter, and a robber" (1180). Blinded and fettered in the "common prison" (1161), "there to grind /Among the slaves and asses" (1161–2), Samson is utterly unremarkable. Once a legendary warrior—"Much I have heard /Of thy prodigious might and feats performed /Incredible to me" (1082–4)—Samson is no longer a marvelous hero of great repute, but is instead a debased slave, unworthy of recognition, a fallen hero whose story no longer matters. Samson's God "Thee he regards not, owns not, hath cut off /Quite from his people" (1157–8); Israel's magistrates "took thee /As a league-breaker" (1184–5) and delivered him "bound /Into our hands" (1184–5). Both Harapha and Israel mistakenly see Samson as a "league-breaker" (1209) who "presumed /Single rebellion and did hostile acts" (1209–10).

Milton, however, suggests that the hero, by refusing Manoa's and Dalila's proffers of nurture, bravely resists reabsorption into the old family/nation. Through his new heroic act of spiritual resistance, Samson redefines his singularity and alienation as signs of his elect status as Israel's redeemer in line with the righteous individualism of Enoch, Noah, and Abraham in *Paradise Lost*: "I was no private but a person raised /With strength suffi-cient and command from Heav'n /To free my country" (1211–13). The story of Samson's legendary feats of strength on behalf of Israel is cut short by his transgressions, which tragically place him outside the Law, his nation, and the heroic tradition. But, his radical break from all that pre-cedes him provides a new beginning and a divinely inspired, rather than merely legendary, heroic plotline. Just as in the invocation to Book 9 of *Paradise Lost*, where Milton condemns Arthurian legend, the "long and tedious havoc" of "fabled knights /In battles feigned" (30–1), as a subject unfit for his post-Reformation, anti-monarchical epic, so in *Samson*, Arthurian legend is associated through the Philistine–Stuart Harapha with the outmoded heroics of physical prowess and imperial conquest. At the end of his visit, the Philistine giant seems less a valiant Lancelot than a cowardly Braggadochio. Samson's reformation from legendary warrior to divinely inspired agent (and victim) of providential history enables the hero to play his role as Israel's emancipator, the role to which he is dedi-cated by angelic promise. Samson's encounter with Harapha inaugurates the narrative of reformed heroism.

As Harapha's visit underscores, Samson's physical prowess ultimately proves to be beside the point; rather "strength sufficient" (1212) is newly identified with Samson's inner resistance to and rejection of both Dalila's and Manoa's proffers of maternal nurture. The dual nature of Samson's rejection is crucial. Not only does the hero's rejection of both Dalila and Manoa "imply the comprehensive discrediting of domesticity because of its danger of eventuating in domination," as Mueller insightfully ob-serves,[47] but it also manifests the submerged identity between the seem-ingly opposed Philistine and Hebrew nations. This collapsing of difference is central to Samson's restitution of his role as Israel's emancipator in in-ternational terms. As already noted, in obliterating the boundary between Philistine and Hebrew, Samson's dual rejection of Dalila and Manoa renders the Judaic division between the uncircumcised and the circum-cised obsolete. But, this rejection also resituates the divide between pure and impure, clean and unclean, within the new inward locus of value that opens up at the missing middle of the drama. In *Paradise Regained*, this inward site is identified with Mary's house. In *Samson*, however, the new

⁴⁷ Janel Mueller, *Dominion as Domesticity*, p. 46.

locus of value is divorced from the private sphere of the maternal-centered home, which is reidentified as the common site of Philistine and Hebrew thralldom. It is mapped instead onto the regenerate international space that transcends the old compromised categories of the public and the private, the paternal and the maternal, the circumcised and the uncircumcised, among other governing binaries. This salvific, neither/nor space comes into view at the middle of the drama, when Samson first says "no" and then says "yes" to the Philistine Officer. This reversal is not a political or religious compromise; nor, finally, is it grounded in the conjunctive, both/and monism that organizes rabbinic Judaism's charitable understanding of the expansiveness of the Law. Rather, the threshold between no and yes that Samson straddles in his dialogue with the Philistine officer provides a portal to the sublime-Pauline paradigm of community as "neither Jew nor Greek," underpinning Milton's new international conception of Protestant community. Neither Dalila's husband nor Manoa's son, Samson, now a free, independent-and-international agent, breaks all natural, blood-and-soil ties to the family-nation of Israel, upon which the drama comprehensively forecloses.[48] He thus preserves and renews his elect status as "separate to God" (32) and God's nursling. Samson's contradictory act of transgression and obedience reflects his new understanding of the equation between God's enforcement and suspension of the law, which aligns highest justice with true liberty. If it leads him away from the maternal-centered home and the old, embodied, or carnal Israel, Samson's reversal from no to yes shows him the way back home in its new de-familiarized worldwide form. Obedience and transgression, separateness and mixture, national and international, provide the mutually constitutive measures of Samson's reformed identity and spiritual freedom.

SEPARATE AND MIXED: SAMSON AND PROTESTANT INTERNATIONALISM

The time and space between Samson's refusal and consent to go with the Philistine messenger highlight his position in the liminal interval after the Law but before grace.[49] But, although superficially the same, this temporal

---

[48] For an opposing interpretation, see Julia Rheinhardt Lupton, "Samson Dagonistes," *Citizen-Saints: Shakespeare and Political Theology* (Chicago: University of Chicago Press, 2005), p. 201. Lupton argues that Samson dies back into kinship and primitive tribalism.

[49] Drawing on Agamben and Schmitt, Victoria Kahn argues that Samson's about-face places the hero in a "state of exception," which perversely confers legitimacy upon his horrible massacre of the Philistines. See "Political Theology and the Realm of the State in *Samson Agonistes*," *South Atlantic Quarterly*, 99.4 (1996): 1066.

space differs considerably from the murky conceptual terrain that Samson occupies in his opening soliloquy.[50] Like the lamenting Adam in Book 10 of *Paradise Lost*, the despairing Samson at the beginning of the drama finds himself in a swarming cognitive state of in-between-ness. In contrast to the physical ease that his body is able to achieve in prison in Gaza, his mind is plagued by "restless thoughts, that like a deadly swarm /Of hornets armed, no sooner found alone /But rush upon me thronging" (19–21). These thoughts make "present /Times past" (22–3), suspending him between "what once I was" and "what am now" (23). Unable to make his birth "from Heaven foretold /Twice by an angel" (24–5) cohere with what he perceives as his impending death. "Betrayed, captive, and both my eyes put out" (33), Samson cannot at first fashion a meaningful narrative of his life. Unable to see his life as the progressive fulfillment of the prophecy announced at his birth, Samson finds neither shape nor order in the profane intermixture of his glorious past and ignoble present. Rather than a sequential and meaningful narrative, Samson's life appears to him as an incoherent and profane mingling of events ordinarily distinct and distant from one another in time and space.[51] Birth and death no longer mark the beginning and end points of his life's narrative, but instead they merge unproductively into an unnarratable state of suspended animation. Time and again, he laments this enthralling, spectral condition of "living death" (100) into which the once expansive and meaningful, narrative scope of his life has been confusedly reduced: "Scarce half I seem to live, dead more than half" (79). "[E]xiled from light" (98), he is compelled to "live a life half dead, a living death, /And buried; but O yet more miserable! /Myself my sepulchre, a moving grave, /Buried, yet not exempt /By privilege of death and burial /From worst of other evils" (100–5). As in Adam's soliloquy, the erasure of distinction between life and death in Samson's lament, as in Jer. 41:5, appears to signal the complete breakdown of the sacred ritual order and the collapse of the separate, sanctified, cultic sphere, rendering circumcision and all other external marks of Hebraic difference and particularity obsolete.

At the beginning of the drama, Samson's violation of Hebraic paradigms of separateness deprives him of the light needed to compose his life into an orderly narrative, consigning him to despair and immobility. But,

---

[50] In "Samson Dagonistes," *Citizen-Saints: Shakespeare and Political Theology* (Chicago: University of Chicago Press, 2005), p. 187, Lupton argues that Samson is reduced to what Giorgio Agamben terms the state of "bare life." Samson subsists not as a political subject, citizen, or resident alien in Gaza, but instead as "a primitive Golem at work in the mills of the military-industrial complex that has swept his primitive strength up into its higher-order function."

[51] See Marshall Grossman, *"Authors to Themselves": Milton and the Revelation of History* (Cambridge: Cambridge University Press, 1987).

after exposing the submerged identity between the Philistine and Hebrew nations by rejecting both Dalila's and Manoa's invitations to nurse him, and after embracing the new heroism and reformed plotline that is set into motion after Harapha's visit, Samson moves from the deadly both/ and condition he describes in his soliloquy to the vital neither/nor site that opens up during his encounter with the Philistine officer. Rather than remain immobilized by his in-between-ness as he is in his opening soliloquy, Samson now gains agency from his neither/nor (neither Hebrew, nor Philistine) state. No longer the site of impasse, his location between the old dispensation and the new one, between the dispensing of and dispensing with the law, represents a reformed, international arena for productive, meaningful, and divinely sanctioned action: "Nothing to do, be sure, that may dishonor /Our law, or stain my vow of Nazarite" (1384–6).

The double-take interval between his initial no and his final yes in his response to the Philistine Officer thus changes (or reforms) change. If earlier in the drama change is equated with misery, "O miserable change!" (340), change is now newly associated with a higher form of continuity. In replacing no for yes, Samson exchanges a system of values that dictates but can also tolerate the differences between circumcised and uncircumcised, clean and unclean, for one that cancels these differences entirely. But Samson perceives this exchange as the equitable conversion of one moral currency for another equally just, but more current, currency. Hence, the drama suggests that, by substituting the former for the latter, Samson loses nothing, but gains everything:

> And for a life who will not change his purpose?
> (So mutable are all the ways of men)
> Yet this be sure, in nothing to comply
> Scandalous or forbidden in our Law.
>
> (1406–9)

Let us note that Samson's focus on change in this passage is nothing new: he has been a changed man from the very start of the play. But, at this pivotal, no-to-yes, moment, Samson revises his perception of change. Whereas at the beginning of the drama Samson's downfall "exile[s] [him] from the light" (98), in his encounter with the Officer, this same ruination liberates his mind and spirit, allowing him to repair the broken narrative of his life. Whereas, after his transgression, Samson can find no coherence amongst his past, present, and future, he now discerns new continuities: that "This day will be remarkable in my life" (1388).

As we have already seen, the lines in Milton's Hellenic-biblical tragedy between a neither/nor and a both/and perspective can seem impossibly confused, and it is Samson's heroic task to make sense of these confusions. In political terms, Samson's hermeneutical *agon* restages seventeenth-century English contests for control of national definition. The struggle between reformed and received, abstract and corporealist, and affective and ethnic models of national community find representation through Samson's struggles with the mind–body, spirit–matter question. In wrestling with these binaries, Samson becomes the vehicle of the play's international vision. His strenuous efforts to reconcile the differences between Pauline dualism and biblical and post-biblical monism bring the separatist discourse of rabbinic Judaism into fruitful proximity with the universalizing, international discourse of Pauline and post-Pauline Christianity. For Milton, this hermeneutical merger provides a crucial intellectual framework for justifying *Samson's* international model of godly community, as at once secessionist and supersessionist, closed and open, exclusive and universal. In converting no into yes, Samson not only finds common ground between old and new, obedience and transgression, the Judaic and the Christian, but in the end, he also converts the former terms into the latter ones.

This epoch-shifting vision of individual and worldwide reformation as coinciding with the conversion and sublation of the Judaic into the Christian finds a crucial, historically specific source in seventeenth-century millenarianism. One fundamental principle of the millenarian movement was the coterminous relations it forged among Jewish conversion, national reform, and world renewal. As Michael Ragussis points out, English millenarian discourse "speaks in the same breath of the Jews' restoration and their conversion." The Jews would be restored to "Israel" (or Israel re-formed at the end of Days) at the precise moment in which they would renounce their Judaism and embrace reformed Christianity.[52] This apocalyptic vision of simultaneous Jewish reclamation and denial, or return and departure, is summed up perfectly by Milton's Jesus in *Paradise Regained*. Jesus prophesies that God "Remembr'ing Abraham, by some wondrous call /May bring [the Jews] back repentant and sincere" (3: 434–5).[53]

---

[52] Michael Ragussis, *Figures of Conversion: "The Jewish Question" and English National Identity* (Durham: Duke University Press, 1996), pp. 91–2.

[53] Milton also alludes to the conversion of the Jews as a sign of the imminence of the Apocalypse in *Observations Upon the Articles of Peace*: "And yet while we detest *Judaism*, we know our selves commanded by St. *Paul, Rom.* 11 to respect the *Jews*, and by all means to endeavour thir conversion (*CPW*, 3:326). He also refers directly to Jewish conversion in *De Doctrina*, where he lists the "calling of the Jews" among other possible signs of the End of Days: "Some authorities think that a further portent will herald this event, namely, the calling of the entire nation not only of the Jews but also of the Israelites." See my " 'The people of Asia and with them the Jews': Asia, Israel, and England in Milton's Writings," *Milton and the Jews*, ed. Douglas Brooks (Cambridge: Cambridge University Press, 2008).

Ragussis notes that "'the call'...that Milton's Christ hesitantly acknowl-edges is a critical touchstone of a tradition of millenarian thought about the relation between the Jews' history and world history."[54] Samson's "change"—his obedience through transgression, his transumptive rela-tion to the law—might be read as a response to this "wond'rous call" (*PR*, 3: 434), one that situates Samson as the converted, "repentant and sincere" (3: 435) Hebrew, who enters world history (the common era) and the universal community by sacrificing his own Hebraic past and particularity.

Much is at stake here, since the "exclusive universalism" inscribed by the Samson's dramatic but undramatizable inner change at the middle of the drama also is implicated in an entangled matrix of ruling and dissent-ing seventeenth-century political and religious discourses. One very com-pelling way to read *Samson Agonistes* is as Milton's revolutionary attempt to wrest control of the Pauline ideal of universal community in Christ from governing anti-tolerationist constructions of England as a nation united spiritually through shared participation in the sacraments, and to make it serve instead nonconformist models of social and religious inclu-siveness. *Samson*, as such, might be said to translate a "ceremonialist" paradigm of community into a "puritan" one, to borrow Achsah Guib-bory's very useful terms. Milton had already made this kind of revisionist move in *Areopagitica* by strategically deploying the reconstituted temple of Solomon, which, as Guibbory notes, "the Laudian Church had in-voked to support uniformity in ceremonial worship,"[55] as an emblem of the unforced gathering of separatist churches, or the inclusiveness pro-claimed by Galatians 3:28. In the reformed Solomonic temple, the Pauline vision of inclusive unity would find expression as the unforced filiations of all sectarian configurations of Christian belief and religious practice brought together in "gracefull symmetry" as "brotherly dissimilitudes" (*CPW*, 2: 555).

The desire to tolerate "brother dissimilitudes" (*CPW*, 2: 555) at home, however, was paralleled by an expansionist ambition to federate all na-tions into a universal brotherhood. For Boyarin, Christian universalism produces "the discourse of conversion, colonialism, the 'white man's burden.'"[56] Those who fail to become brothers become others, to be cast down like Cain, the "other" brother. In *Samson*, the hero points up the close relations between fraternity and fratricide when he destroys the Philistine (br)other along with his Hebrew self as he collapses the pillars

---

[54] Michael Ragussis, *Figures of Conversion: "The Jewish Question" and English National Identity* (Durham: Duke University Press, 1995), p. 92.

[55] Achsah Guibbory, *Ceremony and Community from Herbert to Milton*, pp. 4, 181.

[56] Daniel Boyarin, *Carnal Israel*, pp. 233, 235.

of Dagon's temple. Just as his dual rejection of Dalila's and Manoa's nursing households obliterates the difference between Philistine and Hebrew in the private sphere, so Samson's horrible act of destruction cancels Philistine–Hebrew difference in the public domain. Israel's kinship with the Philistines is at once acknowledged and repudiated, and then refined into a higher Pauline unity, which makes that difference irrelevant.

The link between Samson's physical destructiveness and his dual rejection of Manoa's and Dalila's good and bad nurturing households is made even more precise by the hero's willingness to break "The bond of nature" (*PL*, 9: 956) that connects him exclusively with Israel. Rather than assert his difference as a pure Hebrew from the profane Philistines: "Samson is with these immixt, inevitably" (1657). In breaking his natural, nursing maternal tie to Israel, Samson achieves spiritual independence. He is poised to enter the international fraternity of rational men. The hero's achievement, however, comes at a cost: he must negate all natural forms of intimacy, kinship, and community. Mixed and separate, Samson, through his sacrificial death, becomes the agent of Milton's exclusive-universal vision of Judeo-Christian world history. In collapsing the distinction between the Philistines and himself as a Hebrew, while regaining his separateness to God, Samson restores human experience to its original *tabula rasa* state, before the disintegration of the universal whole into constituencies, competing for preeminence.

The *tabula rasa* metaphor of purity was integral to Puritan ambitions in both the New World abroad and the old one at home.[57] It helped to license the imposition of reformist models of entirely artificial, utopian social organization upon supposedly blank colonial spaces.[58] The Hebraic nation, so crucial to the concerns of *Samson Agonistes*, was England's and Europe's archetypal displaced population. The Jews were forcibly driven not only from ancient Israel, but also again and again from their adopted modern European homelands in Catholic England, Spain, and France. With Protestantism, however, inclusion and conversion provided the reformed means to achieve the ingathering of all peoples into Christian fraternity, which would usher in the Apocalypse. The shift from deportation to toleration, from Expulsion to assimilation, was the key to England's colonial project in the Americas, whose native populations, displaced by European conquest, were refashioned as Jews and assigned Hebraic origins, which were then effaced and subsumed by conversion to

---

[57] James Holstun, *A Rational Millennium: Puritan Utopias of Seventeenth-Century England and America* (New York: Oxford University Press, 1987), pp. 10, 105.
[58] Jonathan Boyarin, *Storm From Paradise: The Politics of Jewish Memory* (Minneapolis: University of Minnesota Press, 1992), p. 78.

Christianity; this is a double conversionary process.[59] As Bruce McLeod observes, militant Protestants like Milton "saw the 'nation's apocalyptic renewal' as inseparable from the rejuvenation of ungodly nations and peoples the world over."[60] It is important to note in this context that theories about the Hebraic origins of the American Indians reached full force at the precise moment in which Readmission was being fiercely debated. The Judaizing of the New World and the Readmission of the Jews to a tolerant new England are energized by the same world vision. As Holstun notes, seventeenth-century millenarianism posits "an ultimate (or penultimate) unity of peoples and of history"—of a world population descended from the dispersed or lost tribes of Israel, all awaiting conversion and ingathered community at the Apocalypse.[61] Within this visionary framework, justifiable, and so-called humanitarian acts of conquest and conversion, such as Cromwell's brutal invasion of Ireland and his war with Spain over Hispaniola, in conjunction with tolerationist policies such as the Readmission Act, would help to usher in the global Protestantism to be achieved at world history's salvific endpoint. This End would also coincide with the Beginning, since the apocalyptic unifying and conversion of all peoples, all descended from Israel, into reformed Christian federation restores the world to its pure, seamless originary moment by erasing all precursor histories, synthesized as Hebraic.

In *Samson Agonistes*, this same violent, conversionary link between pure origin and apocalyptic ending is represented by Samson's phoenix-like integration of death and rebirth at the close. As the allegorical phoenix, Samson's pointlessly destructive actual death is purified by spiritual fire into a newly meaningful, abstract beginning. Although, in literal terms, Samson's devastating acts of violence accomplish nothing (Israel remains under Philistine bondage), in allegorical terms, the same violence emancipates Samson and Israel from the embodied and engulfing Hebraic past, which is associated with the nursing mother. Subsumed and then transumed in the androgynous phoenix's "ashy womb" (1703), the hero is reborn in spirit as God's nursling. In *Paradise Lost*, these very same regenerative ashes are associated with the "New Heav'n and Earth" to be ush-

[59] On the Hebraic origins of the American Indians, see Lee Eldridge Huddleston, *Origins of the American Indians: European Concepts, 1492–1729* (Austin: University of Texas Press, 1967); Richard H. Popkin, "Jewish Messianism and Christian Millenarianism," in *Culture and Politics from Puritanism to the Enlightenment*, ed. Perez Zagorin (Berkeley and Los Angeles: University of California Press, 1980), pp. 67–90; David S. Katz, "The Debate over the Lost Ten Tribes of Israel," *Philo-Semitism and the Readmission of the Jews to England 1603–1655* (Oxford: Clarendon Press, 1982), Chapter 4.

[60] Bruce McLeod, "The 'Lordly Eye': Milton and the Strategic Geography of Empire," in *Milton and the Imperial Vision*, p. 51.

[61] James Holstun, *A Rational Millennium*, p. 113.

ered in at the End of Days: "Meanwhile /The world shall burn, and from her ashes spring /New Heav'n and Earth, wherein the just shall dwell" (3: 334–5). This consummation is both universal, creating a world in which "God shall be all in all" (3: 341), and exclusive, "wherein [only] the just shall dwell" (3: 335). The gap between the old world and the "New Heav'n and Earth" (3: 335) also traverses the divide between the "carnal Israel" alluded to in 1 Corinthians 10:18 ("Behold Israel according to the flesh") and the spiritual Israel heralded by the gospels and Paul's epistles. For unlike the old Israel, in which identity is rooted in the body and closely associated with maternal nurture, the new Israel is an international community of spiritual converts (for whom the distinctness of national identity is irrelevant). Through his exemplary role as a repentant and, ultimately, de-particularized or "ashy" Hebrew, Samson-as-phoenix allegorizes the exclusive-universal Israel.

On the one hand, Samson's phoenix-like rebirth reattaches him to God's breast. On the other hand, Samson, like "that self-begott'n bird," (1699), also becomes his own author—hence his insistence that his final act will be "of my own accord" (1643).[62] The coincidence between Samson's return to the deity's nurturing breast and his final assertion of personal autonomy raise two interlocking political points. First, by attaching himself exclusively to God, Samson radically separates himself from all embodied-and-natural forms of human belonging. Even in their most charitable and nurturing configurations, Samson's natural-and-native bonds to father, wife, and country are constraining. This is why his dual rejection of Manoa and Dalila as maternal nurses is crucial. Second, by repudiating these natural, maternal bonds, the newly foreign-autonomous Samson is paradoxically poised to enter the de-individuated sphere of international Protestantism. Incorporating his reborn self with regenerate selves the world over, Samson transcends the embodied particularities of ethnicity, class, gender, and history.

Samson succeeds where Adam fails. By resisting the "bond of nature" (*PL*, 9: 956), Samson sacrifices all local and particular forms of familial, spousal, and national identity for spiritual membership in the universal family of nations. But, in so doing he acquires the will to act on his own accord, and he thereby achieves the liberty of conscience that for Milton characterizes the reformed subject. Adam's fall follows upon his failure to make this same sacrifice, to ignore his natural bond to Eve, and to act with autonomy or free will. Samson's terrible sacrifice models the sacrifice that Milton suggests all men must make in order to reverse the effects of

---

[62] See Grossman, *"Authors to Themselves": Milton and the Revelation of History* (Cambridge, 1987).

the Fall, to regain their native liberties in spiritual terms, and to renew the global promise of a godly community. Like the terribly flawed Samson, fallen humankind must summon up the incredible spiritual strength required to give up inborn, genealogical measures of elect national identity and, instead, to strive for incorporation within the international, spiritual body-politic of the godly. To break these natural bonds, the regenerate individual must resist the almost irresistible pull of the embodied love and nurture that intimately bring parents and children, husbands and wives, together as "One flesh" (*PL*, 9: 959). By performing this powerful inward act of resistance, Samson allegorically fulfills his role as Israel's emancipator, despite his literal failure to liberate his nation from Philistine oppression.

Although largely ignored by readers of Milton's complex tragedy, *Samson*'s foreclosure on the nursing maternal-centered realm as a regenerative, alternative domain is crucially linked to the drama's vision of worldwide Protestant reformation. In rejecting both Dalila's and Manoa's offers to nurse him, Samson not only collapses the national/ethnic difference between the Philistines and the Hebrews, but he also resists the almost irresistible maternal pull of natural kinship and the family/nation. "Neither Jew nor Greek," Manoa's son nor Dalila's husband, the reformed Samson exemplifies both the universal Christian subject and the reformed Protestant individual. Breaking "The bond of nature" (*PL*, 9: 956) exacts a terrible price: isolation, alienation, exile, historical erasure, destruction, and death. But, for Milton, this is a price that all men must be willing to pay in order for communion and community with God and the godly.

Finally, in repudiating the reformed model of nurturing household government that serves the very different nationalist interests of the wide-ranging texts already considered, *Samson Agonistes* opens up new fault lines in the historical and cultural moment that my book interrogates. In *Samson*, Milton represents the nurturing household's potential for reform as impossibly compromised. Unlike the other texts I have examined, Milton's drama dissociates the reformation of the nation from the invention of motherhood and the maternal-centered household. Rather than elevate the figure of the nursing mother into an emblem of national unity and renewal, Samson's dual rejection of Manoa's and Dalila's nursing not only forecloses on the domestic sphere, but also resituates reformation in the international incorporation of regenerate souls. Although Milton does not entirely give up on the redemptive power of mother-love and nurture, he divorces such power from the new mother, even in her most radically reformed, spiritually purified guise, as found in the guidebooks, the gender-debate pamphlets, and in Milton's own prose tracts, *Paradise Lost* and *Paradise Regained*. For this reason, the poet suppresses Samson's mother and the Hebraic/Puritan linkage between pure nurture and pure national identity. With the possible exception of "Mary, second Eve" (*PL*, 10: 183), a unique example of maternal perfection in human form, Milton attributes maternal-love and nurture to the deity alone. Despite its profound sympathies for and deep understanding of rabbinic Judaism's expansive interpretation of the Law, *Samson Agonistes* suggests that, in the end, the paradoxical mixture of connectedness and autonomy, universalism and election, which defines the regenerate Protestant subject, can be realized only by rejecting the Law's maternal measures for constituting national identity. Samson must turn away from his old nurturing mothers, Dalila, Manoa, and the nation of Israel/England, so that he can free himself completely from outmoded land, blood, and milk measures of pure identity, even in their most charitable forms. By exchanging his natural mothers and fathers for the nurturing deity and translating Israel into "Israel," he establishes "what the reformers describe [as] ... the coincidence between right knowledge of self and right knowledge of God."[63] With his emancipated will and renewed dependence on the deity, Samson regains his special distinction as God's nursling: in spirit, he becomes a "member incorporate" of the international community of the godly.

---

[63] Victoria Silver, *Imperfect Sense: The Predicament of Milton's Irony* (Princeton: Princeton University Press, 2001), p. 246.

Bibliography

Abbott, Wilber Cortez, ed. *The Writings and Speeches of Oliver Cromwell.* 4 vols. Cambridge, Mass: Harvard University Press, 1937.

Achinstein, Sharon. "The Politics of Babel in the English Revolution." In *Pamphlet Wars: Prose in the English Revolution*, ed. James Holstun, pp. 14–44. London: Frank Cass, 1992.

———. *Milton and the Revolutionary Reader.* Princeton: Princeton University Press, 1994.

———. "Women on Top in the Pamphlet Literature of the English Revolution." *Women's Studies* (1994): 131–64.

———. *Literature and Dissent in Milton's England.* Cambridge: Cambridge University Press, 2003.

Adelman, Janet. *Suffocating Mothers: Fantasies of Maternal Origin in Shakespeare's Plays, Hamlet to The Tempest.* New York and London: Routledge, 1992.

A.M. *A Rich Closet of Physical Secrets... viz. The Child-Bearers Cabinet.* London: Gartrude Dawson, 1652. Thomason/E. 670[1].

Anderson, Benedict. *Imagined Communities: Reflections on the Origin and Spread of Nationalism.* London: Verso, 1983.

Anon. *The Court and Kitchen of Elizabeth.* London: Thomas Milbourne, 1664. Wing/C7636.

Archer, John Michael. *Citizen Shakespeare: Freemen and Aliens in the Language of the Plays.* New York: Palgrave-Macmillan, 2005.

Ariès, Phillipe. *Centuries of Childhood: A Social History of Family Life.* Trans. Robert Baldrick, London: John Cape Ltd, 1962.

Augustine. *Tractatus adversus Judaeos.* In *Fathers of the Church.* Trans. M. Ligouri. Washington, DC: Catholic University of America Press, 1955.

Austern, Linda. "'My Mother Musicke': Music and Early Modern Fantasies of Embodiment". In *Maternal Measures*, eds. Naomi J. Miller and Naomi Yavneh, pp. 239–81. Burlington, VT: Ashgate, 2000.

Aylmer, G.E. *Rebellion or Revolution?: England 1640–60.* Oxford: Oxford University Press, 1986.

Bacon, Francis. *A Brief Discourse Touching the Happy Union of the Kingdomes of England and Scotland.* London: [R. Read], 1603. STC (2nd ed.)/1117.

Bacon, Francis. *Francis Bacon, The Works of Francis Bacon.* 14 vols. Ed. James Spedding, Robert Leslie Ellis, and Douglas Denon Heath. London: Longman, 1860.

Baskins, Cristelle L. "Typology, Sexuality, and the Renaissance Esther." In *Sexuality and Gender in Early Modern Europe: Institutions, Texts, Images*, ed. James Grantham Turner, pp. 31–54. Cambridge: Cambridge University Press, 1993.

Baston, Jane. "History, Prophecy, and Interpretation: Mary Cary and Fifth Monarchism." *Prose Studies*, 21.3 (December 1998): 1–18.

Berry, Boyd M. "The First English Pediatricians and Tudor Attitudes Toward Childhood." *Journal of the History of Ideas*, 35.4 (1974): 561–77.

Borin, Françoise. "Judging By Images." In *A History of Women in the West*, eds. Natalie Zemon Davis and Arlette Farge, pp. 187–254. Cambridge, MA: Belknap-Harvard University Press, 1993.

Boyarin, Daniel. *Carnal Israel: Reading Sex in Talmudic Culture*. Berkeley: University of California Press, 1993.

——. "'These and These are the Words of the Living God': Apophatic Rabbinism," paper, Dartmouth Regional Seminar in Jewish Studies, Hanover, NH, May 2002.

Boyarin, Jonathan. *Storm From Paradise: The Politics of Jewish Memory*. Minneapolis: University of Minnesota Press, 1992.

Brathwaite, Richard. *The English Gentleman and English Gentlewoman*. Wing/ B4262, 1641, Frontispiece.

Bulwer, John. *Anthropometamorphosis: Man Tranform'd*. 2nd ed. London: J. Hardesty, 1653.

Burt, Richard and John Michael Archer, eds. *Enclosure Acts: Sexuality, Property, and Culture in Early Modern England*. Ithaca: Cornell University Press, 1996.

Bynum, Caroline Walker. *Holy Feast and Holy Fast: The Religious Significance of Food to Medieval Women*. Berkeley: University of California Press, 1987.

——. *Jesus as Mother: Studies in the Spirituality of the High Middle Ages*. Berkeley, Los Angeles, and London: University of California Press, 1982.

Callaghan, Dympna. "Wicked Women in *Macbeth*: A Study of Power, Ideology, and the Production of Motherhood." In *Reconsidering the Renaissance*, ed. Mario A. Di Cesare, pp. 355–69. Birmingham, NY: Medieval and Renaissance Texts and Studies, 1992.

Calvin, John. *Tracts and Treatises on the Reformation of the Church*. Trans. Henry Beveridge. 3 vols. Grand Rapids: William B. Eerdman Publishing Co., 1958.

Camden, William. *Britannia: Or a Chorographicall description of the flourishing Kingdoms of England, Scotland, and Ireland...* (1607), with an English translation by Philemon Holland. Hypertext edition by Dana F. Sutton. University of California, Irvine, 2004.

Cary, Mary. *The little horns doom & downfall or A scripture-prophesie of King James and King Charles, and of the present Parliament*. Thomason/E.1274(1). London, 1651.

Chamberlen, Stephanie. "Fantasizing Infanticide: Lady Macbeth and the Murdering Mother in Early Modern England." *College Literature*, 82.3 (2005): 72–91.

Cheney, Patrick. "'O let my books be...dumb presages': Poetry and Theatre in Shakespeare's Sonnets." *Shakespeare Quarterly*, 52.2 (Summer 2001): 222–54.

Clark, Alice. *Working Life of Women in the Seventeenth Century*. 1919; reprint, London: Routledge and Kegan Paul, 1982.

Cleaver, Robert and John Dod. *A Briefe Explanation of the Whole Booke of the Proverbs of Salomon*. London: Felix Kyngston, 1615. STC (2nd ed.) / 5378.1.

Cleaver, Robert and John Dod. *A Godly Form of Household Government: For the Ordering of Private Families, according to the Direction of God's Word.* London: R Field, 1621. STC (2nd ed.) / 5387.5.

Clinton, Elizabeth. *The Countess of Lincolnes Nurserie.* Oxford, 1622.

Colley, Linda. *Britons: Forging the Nation, 1707–1837.* New Haven: Yale University Press, 1992.

Coiro, Ann Baynes. "Milton & Sons: The Family Business." *Milton Studies*, 48 (2008): 13–37.

Comenius, John Amos. *The Great Didactic of John Amos Comenius.* Trans. M.W. Keating. New York: Russell and Russell, 1967.

Cook, John. *King Charls his case: or, An appeal to all rational men, concerning his trial at the High Court of Justice, 1649.* Thomason/84:E.542(3). London, 1649.

Cope, Abiezer. *A Second Fiery Flying Roll.* London, 1650.

Corns, Thomas. "Duke, Prince and King." In *The Royal Image: Representations of Charles I*, ed. Thomas N. Corns, pp. 1–25. Cambridge: Cambridge University Press, 1999.

Correll, Barbara. "Malleable Material, Models of Power: Women in Erasmus's 'Marriage Group' and *Civility in Boys.*" *ELH*, 57 (1990): 241–62.

Cotton, John. *A Briefe Exposition of the whole Book of Canticles...* London: Philip Nevil, 1642. Wing/C6410. *EEBO.*

Coward, Barry. *Oliver Cromwell.* London: Longman, 1991.

Crawford, Julie. *Marvelous Protestantism: Monstrous Births in Post-Reformation England.* Baltimore: Johns Hopkins University Press, 2005.

Cressy, David. *Birth, Marriage, and Death: Ritual, Religion, and the Life-Cycle in Tudor and Stuart England.* Oxford: Oxford University Press, 1999.

Crooke, Helkiah. *Microcosmographia: A Description of the Body of Man.* London: William Iaggard, 1615. STC/6062.2. *EEBO.*

Cross, Claire. "The Church in England 1646–1660." In G.E. Aylmer, ed. *The Interregnum: The Quest for Settlement.* Hamden, CT: Archon-Shoe String, 1972.

Cust, Robert and Ann Hughes, eds. *Conflict in Early Stuart England: Studies in Religion and Politics, 1603–42.* London and New York: Longman, 1989.

Davies, Lady Eleanor Davies. *A Warning to the Dragon and all his Angels* (1625). In *Prophetic Writings of Lady Eleanor Davies*, ed. Esther S. Cope, pp. 1–56. Oxford: Oxford University Press, 1995.

Davies, Stevie. *Unbridled Spirits: Women of the English Revolution, 1640–1660.* London: The Women's Press, 1998.

Davis, J.C. "Cromwell's Religion." In *Oliver Cromwell and the English Revolution*, ed. John Morrill, pp. 181–208. London: Longman, 1990.

de Montaigne, Michel. *The Complete Works of Montaigne.* Ed. and trans. Donald M. Frame. Palo Alto: Stanford University Press, 1956.

Dolan, Frances. "Marian Devotion and Maternal Authority in Seventeenth-Century England." In *Maternal Measures*, eds. Naomi J. Miller and Naomi Yavneh, pp. 282–91. Burlington: Ashgate, 2000.

Donne, John. *The Complete Poetry and Selected Prose of John Donne.* Ed. Charles M. Coffin. New York: Modern Library, 1952.

Donnelly, M.L. " 'And still new stopps to various time apply'd': Marvell, Cromwell, and the Problem of Representation at Midcentury." In *On the Celebrated and Neglected Poems of Andrew Marvell*, eds. Claude J. Summers and Ted-Larry Pebworth, pp. 163–76. Columbia: University of Missouri Press, 1988.

Drayton, Michael. *Poly-Olbion*. London, 1612. Frontispiece.

Dubrow, Heather. *A Happier Eden: The Politics of Marriage in the Stuart Epithalamium*. Ithaca: Cornell University Press, 1990.

Eagleton, Terry. *William Shakespeare*. Oxford: Blackwell, 1986.

Eisenstein, Elizabeth. *The Printing Press as an Agent of Change*. Cambridge: Cambridge University Press, 1985.

Elliot, Dylan. *Fallen Bodies: Pollution, Sexuality, and Demonology in the Middle Ages*. Philadelphia: University of Pennsylvania Press, 1999.

Erasmus, Desiderius. *The Colloquies of Erasmus*. Trans. Craig R. Thompson. Chicago: University of Chicago Press, 1965.

Erickson, Amy Louise. *Women and Property in Early Modern England*. London: Routledge, 1993.

Evans, James Martin. *Milton's Imperial Epic: "Paradise Lost" and the Discourse of Colonialism*. Ithaca: Cornell University Press, 1996.

Ezell, Margaret J.M. *The Patriarch's Wife: Literary Evidence and the History of the Family*. Chapel Hill: University of North Carolina Press, 1987.

Fildes, Valerie A. *Breasts, Bottles, and Babies: A History of Infant-Feeding*. Edinburgh: University of Edinburgh Press, 1986.

——. *Wet-Nursing: A History from Antiquity to the Present*. Oxford: Basil Blackwell, 1988.

Filmer, Robert Sir. *Patriarcha: Or the Natural Power of Kings*. London: Walter Davis, 1680. Wing/F922. *EEBO*.

Firth, C.H. *Oliver Cromwell and the Rule of Puritans in England*. London: Putnam, 1901.

Fish, Stanley. "Spectacle and Evidence in *Samson Agonistes*." *Critical Inquiry*, 15 (1989): 556–86.

Fliegelman, Jay. *Prodigals and Pilgrims: The American Revolution against Patriarchal Authority, 1750–1800*. Cambridge: Cambridge University Press, 1982.

Floyd-Wilson, Mary. "English Epicures and Scottish Witches." *Shakespeare Quarterly*, 57.2 (Summer 2006): 161–91.

Foucault, Michel. *The Birth of the Clinic: Archaeology of Medical Perception*. Trans. Alan Sheridan. London: Tavistock, 1976.

Fox, Margaret Askew Fell. *A Call to the Universal Seed of God*. London, 1665. Wing/F625A. *EEBO*.

Gardiner, S.R. *History of the Commonwealth and Protectorate*. 2 vols. London: Longmans, Green, & Co, 1903.

Gaunt, Peter. *Oliver Cromwell*. Oxford: Blackwell, 1996.

Gillespie, Katharine. *Domesticity and Dissent in the Seventeenth Century: English Women Writers and the Public Sphere*. Cambridge: Cambridge University Press, 2004.

——. "Elizabeth Cromwell's Kitchen Court: Republicanism and the Consort." *Genders Online*, 33 (2001).

Goldberg, Jonathan. *James I and the Politics of Literature*. Baltimore: Johns Hopkins University Press, 1983.

Goreau, Angeline. *The Whole Duty of a Woman: Female Writers in Seventeenth-Century England*. Garden City, NY: Doubleday, 1985.

Gouge, William. *Of Domesticall Duties. Eight Treatises*. London: John Haviland, 1622. STC (2nd ed.)/12119. *EEBO*.

Greenblatt, Stephen. *Renaissance Self-Fashioning: From More to Shakespeare*. Chicago: University of Chicago Press, 1980.

——. *Shakespearean Negotiations: The Circulation of Social Energy in Renaissance England*. Oxford: Clarendon Press, 1988.

——. "Mutilation as Meaning." In *The Body in Parts: Fantasies of Corporeality in Early Modern Europe*, eds. Carla Mazzio and David Hillman, pp. 221–43. London: Routledge, 1997.

Greenblatt, Stephen, Walter Cohen, Jean E. Howard, and Katharine Eisaman Maus, eds. *The Norton Shakespeare*. New York: W.W. Norton, 1997.

Gregerson, Linda. *The Reformation of the Subject: Spenser, Milton, and the English Protestant Epic*. Cambridge: Cambridge University Press, 1995.

——. "Colonials Write the Nation: Spenser, Milton, and England on the Margins." In *Milton and the Imperial Vision*, eds. Balachandra Rajan and Elizabeth Sauer, pp. 169–90. Pittsburgh: Duquesne University Press, 1999.

Grossman, Marshall. *"Authors to Themselves": Milton and the Revelation of History*. Cambridge: Cambridge University Press, 1987.

Grotius, Hugo. *The Rights of War and Peace, including the Law of Nature and of Nations*. Trans. Richard Tuck. Ithaca: Cornell University Press, 2009.

Guibbory, Achsah. *Ceremony and Community from Herbert to Milton: Literature, Religion, and Cultural Context in Seventeenth-Century England*. Cambridge: Cambridge University Press, 1998.

——. "'The Jewish Question' and 'The Woman Question' in *Samson Agonistes*: Gender, Religion, and Nation. In *Milton and Gender*, ed. Catherine Gimelli Martin, pp.184–206. Cambridge: Cambridge UP, 2004.

——. *Christian Identity, Jews, and Israel in Seventeenth-Century England*. Oxford: Oxford University Press, 2010.

Guillemeau, James (Jacques). *The Nursing of Children. Wherein is set downe, the ordering and gouernment of them from their birth*, affixed to *Childbirth, or the Happie Deliverie of Women*. London: A. Hatfield, 1612. STC (2nd ed.)/ 12496.

Guillory, John. "The Father's House: *Samson Agonistes* in its Historical Moment." In *Re-membering Milton: Essays on the Texts and Traditions*, eds. Mary Nyquist and Margaret W. Ferguson, pp. 148–78. New York and London: Methuen, 1987.

Hall, Joseph. *The Discovery of the New World, or, A Description of the South Indies. Hetherto vnknowne / by an English mercury*. Trans. John Healy. London: Eds. Blount and W. Barrett, 1609. STC (2nd ed.) /12686.

Hall, Kim F. *Things of Darkness: Economies of Race and Gender in Early Modern England*. Ithaca: Cornell University Press, 1995.

——. "Culinary Spaces, Colonial Spaces: The Gendering of Sugar in the Seventeenth Century." In *Feminist Readings of Early Modern Culture: Emerging Subjects*, eds. Valerie Traub, M. Lindsay Kaplan, and Dympna Callaghan, pp. 148–76. Cambridge: Cambridge University Press, 1996.

Harrington, James. *Oceana. The Political Works of James Harrington*. Ed. J.G.A. Pocock. Cambridge: Cambridge University Press, 1977.

Harris, Jonathan Gil. "This is Not a Pipe: Water Supply, Incontinent Sources, and the Leaky Body Politic." In *Enclosure Acts: Sexuality, Property, and Culture in Early Modern England*, eds. Richard Burt and John Michael Archer, pp. 203–28, 1996.

Harvey, William. *Anatomical Excercitations, Concerning the Generation of Living Creatures: To Which are added Particular Discourses, of Births, and of Conceptions, &c*. London: James Young, 1653. Wing/H1085. *EEBO*.

Haskin, Dayton. *Milton and the Burden of Interpretation*. Philadelphia: University of Pennsylvania Press, 1994.

Healy, Simon. "Perceval, Richard (*c*. 1588–1620)". *Oxford Dictionary of National Biography*, online edition, Oxford University Press, September. 2006: http://www.oxforddnb.com/view/article/21914.

Heinemann, Margot. *Puritanism and Theatre: Thomas Middleton and Opposition Drama under the Early Stuarts*. Cambridge: Cambridge University Press, 1980.

Helgerson, Richard. *Forms of Nationhood: The Elizabethan Writing of England*. Chicago: University of Chicago Press, 1992.

Henderson, Katherine Usher and Barbara F. McManus, eds. *Half Humankind: Contexts and Texts of the Controversy about Women in England, 1540–1640*. Urbana and Chicago: University of Illinois Press, 1985.

Herbert, George. *The Works of George Herbert*. Ed. F.E. Hutchinson. Oxford: Clarendon Press, 1941.

Hessup, Cynthia. "The King's Two Genders." *The Journal of British Studies*, 45.3 (July 2006): 493–510.

Heywood, Oliver. *Diaries of Oliver Heywood*. Ed. J.H. Turner. 4 vols. London: Brighouse, 1883.

Hieron, Samuel. *A Help vnto Deuotion: Containing Certain Moulds or Forms of Prayer*. STC 13407. London, 1621.

Hill, Christopher. *Puritanism and Revolution: Studies in Interpretation of the English Revolution of the Seventeenth Century*. London: Secker, 1958.

——. *Reformation to Industrial Revolution: The Making of Modern English Society, 1530–1780*. New York: Pantheon, 1968.

——. *The World Turned Upside Down: Radical Ideas during the English Revolution*. New York: Viking, 1972.

——. "William Harvey and the Idea of Monarchy." In *The Intellectual Revolution of the Seventeenth Century*, ed. Charles Webster, pp. 160–81. London: Routledge, 1974.

——. *Collected Essays II: Religion and Politics in Seventeenth-Century England*. Brighton: Harvester, 1986.

——. *God's Englishman: Oliver Cromwell and the English Revolution*. New York: Penguin Books, 1990.

Hinds, Hilary. *God's Englishwomen: Seventeenth-Century Radical Sectarian Writing and Feminist Criticism*. Manchester: University of Manchester Press, 1996.

Hirst, Derek. *Authority and Conflict: England 1603–1658*. London: Edward Arnold, 1986.

Holinshed, Raphael. *Raphael Holinshed, Chronicles*. Ed. E. G. Boswell-Stone. London: Lawrence, 1986.

Holstun, James. *A Rational Millennium: Puritan Utopias of Seventeenth-Century England and America*. New York: Oxford University Press, 1987.

——. *Ehud's Dagger: Class Struggle in the English Revolution*. London: Verso, 2000.

Hooker, Thomas. *The soules vocation or effectual calling to Christ*. London: John Haviland, 1638. STC (2nd ed.)/13739. *EEBO.*

Howard, Jean E. "Crossdressing, the Theatre, and Gender Struggle in Early Modern England." *Shakespeare Quarterly*, 39.4 (Winter 1988): 418–40.

Howard, Jean E. and Phyllis Rackin. *Engendering A Nation: A Feminist Account of Shakespeare's English Histories*. London and New York: Routledge, 1997.

Huddleston, Lee Eldridge. *Origins of the American Indians: European Concepts, 1492–1729*. Austin: University of Texas Press, 1967.

Hutchinson, Lucy. *Memoirs of the Life of Colonel Hutchinson*. Ed. James Sutherland. Oxford: Oxford University Press, 1973.

Illick, Joseph E. "Child-Rearing in Seventeenth-Century America." In *The History of Childhood*, ed. Lloyd de Mause, pp. 303–50. New York: The Psychohistory Press, 1974.

Jones, Ann Rosalind and Peter Stallybrass. *Renaissance Clothing and the Materials of Memory*. Cambridge: Cambridge University Press, 2000.

Jones, Ernest. *The Life and Work of Sigmund Freud*. Harmondsworth: Penguin Books, 1964.

Jones, John. *The Arte and Science of preseruing Bodie and Soule in all healthe, Wisedome, and Catholike Religion*. London: Henrie Bynneman, 1579. STC (2nd ed.)/14724. *EEBO.*

Jonson, Ben. *Pleasure Reconciled to Vertue*. In *Ben Jonson*. Ed. C. H. Herford and Percy and Evelyn Simpson. 11 vols. Oxford: Clarendon, 1941.

Josephus, *The Works of Josephus*. Trans. William Whiston. Peabody, Mass: Hendrickson Publishers, 1987.

Kahn, Coppelia. "The Absent Mother in *King Lear*." In *Rewriting the Renaissance: The Discourses of Sexual Difference in Early Modern Europe*, eds. Margaret W. Ferguson, Maureen Quilligan, and Nancy J. Vickers, pp. 33–49. Chicago: University of Chicago Press, 1986.

Kahn, Victoria. "Political Theology and Reason of State in *Samson Agonistes*." *South Atlantic Quarterly*, 99.4 (1996): 1065–97.

——. "Disappointed Nationalism: Milton in the Context of Seventeenth-Century Debates about the Nation-State." In *Early Modern Nationalism and Milton's England*, eds. David Loewenstein and Paul Stevens, pp. 249–72. Toronto: University of Toronto Press, 2008.

Kahr, Madlyn. "Delilah." *The Art Bulletin*, 54.3 (September 1972): 282–99.

Katz, David S. *Philo-Semitism and the Readmission of the Jew to England, 1603–1655.* Oxford: Clarendon Press, 1982.

Kelly, Joan. "Did Women Have a Renaissance?" In *Women, History, and Theory: The Essays of Joan Kelly*, pp. 19–50. Chicago: University of Chicago Press, 1984.

Kidd, Collin. *British Identities Before Nationalism: Ethnicities and Nationhood in the Atlantic World, 1600–1800.* Cambridge: Cambridge University Press, 1999.

King, Helen, "The Chamberlens." In *Oxford Dictionary of National Biography*, online ed, Oxford University Press, September 2004: http://www.oxfordddnb.com/view/article15062.

———. *Midwifery, Obstetrics, and the Rise of Gynaecology: The Uses of a Sixteenth-Century Compendium.* Burlington, VT: Ashgate, 2007.

Klein, Melanie. *The Selected Melanie Klein.* Ed. Juliet Mitchell. New York: Free Press, 1986.

Knachel, Philip A., ed. *Eikon Basilike: The Portraiture of His Majesty in His Solitudes and Sufferings.* Ithaca: Cornell University Press, 1966.

Knafla, Louis A. *Law and Politics in Jacobean England: The Tracts of Lord Chancellor Ellesmere.* Cambridge: Cambridge University Press, 1977.

Knoppers, Laura Lunger. *Historicizing Milton: Spectacle, Power, and Poetry in Restoration England.* Athens: University of Georgia Press, 1994.

———. *Constructing Cromwell: Ceremony, Portrait, and Print, 1645–1661.* Cambridge: Cambridge University Press, 2000.

———. "Opening the Queen's Closet: Henrietta Maria, Elizabeth Cromwell, and the Politics of Cookery." *Renaissance Quarterly*, 60.2 (Summer 2007): 464–99.

———. "Consuming Nations: Milton and Luxury." In *Early Modern Nationalism and Milton's England*, eds. David Loewenstein and Paul Stevens, pp. 331–55. Toronto: University of Toronto Press, 2008.

Knott, John. "'Suffering for Truths Sake': Milton and Martyrdom." In *Politics, Poetics, and Hermeneutics in Milton's Prose*, eds. David Loewenstein and James Turner, pp. 153–70. Cambridge: Cambridge University Press, 1990.

Korda, Natasha. *Shakespeare's Domestic Economies: Gender and Property in Early Modern England.* Philadelphia: University of Pennsylvania Press, 2002.

Kowaleski-Wallace, Elizabeth. *Consuming Subject: Women, Shopping, and Business in the Eighteenth Century.* New York: Columbia University Press, 1997.

Krier, Theresa M. *Birth Passage: Maternity and Nostalgia, Antiquity to Shakespeare.* Ithaca: Cornell University Press, 2001.

Labriola, Albert C. "The Aesthetics of Self-Diminution: Christian Iconography and *Paradise Lost*." *Milton Studies*, 7 (1975): 267–311.

Laqueur, Thomas A. *Making Sex: Body and Gender from the Greeks to Freud.* Cambridge: Cambridge University Press, 1990.

Leverenz, David. *The Language of Puritan Feeling: An Exploration in Literature, Psychology, and Social History.* New Brunswick, NJ: Rutgers University Press, 1980.

Levin, William R. *The Allegory of Mercy at the Misericordia in Florence: Historiography, Context, Iconography, and the Documentation of Confraternal Charity in the Trecento.* Lapham, MD: University Press of America, 2004.

Lewalski, Barbara Kiefer. *Paradise Lost and the Rhetoric of Literary Forms.* Princeton: Princeton University Press, 1985.

———. *The Life of John Milton.* Oxford: Blackwell, 2003.

Lieb, Michael. "'Our Living Dread': The God of *Samson Agonistes.*" *Milton Studies*, 33 (1997): 3–25.

———. "'A Thousand Fore-Skins': Circumcision, Violence, and Selfhood in Milton." *Milton Studies*, 38 (2000): 198–219.

Lim, Richard. *Public Disputation: Power and Social Order in Late Antiquity.* Berkeley and Los Angeles: University of California Press, 1994.

Lindley, David. *The Trials of Frances Howard: Fact and Fiction at the Court of King James.* New York: Routledge, 1993.

Loewenstein, David. *Milton and the Drama of History: Historical Vision, Iconoclasm, and the Literary Imagination.* Cambridge: Cambridge University Press, 1990.

———. "The King Among the Radicals: Godly Republicans, Levellers, Diggers and Fifth Monarchists." In *The Royal Image: Representations of Charles I*, ed. Thomas N. Corns, pp. 96–122. Cambridge: Cambridge University Press, 1999.

———. *Representing Revolution in Milton and His Contemporaries: Religion, Politics, and Polemics in Radical Protestantism.* Cambridge: Cambridge University Press, 2001.

Lupton, Julia Reinhardt. *Citizen-Saints: Shakespeare and Political Theology.* Chicago: University of Chicago Press, 2005.

Luther, Martin. *Luther's Works.* Volume 1: Lectures of Genesis 1–5. Ed. Jaroslav Pelikan. St. Louis: Concordia Publishing House, 1958; rpt. 1995.

Luxon, Thomas. *Literal Figures: Puritan Allegory and the Reformation Crisis in Representation.* Chicago: University of Chicago Press, 1995.

Lyly, John. *Euphues* (1580). *The Complete Works of John Lely.* 2 vols. Ed. R. Warwick Bond. Oxford: Clarendon Press, 1902).

Macfarlane, Alan. *Marriage and Love in England: Modes of Reproduction, 1300–1840.* Oxford: Oxford University Press, 1986.

Mack, Phyllis. "Women as Prophets During the English Civil War." *Feminist Studies*, 8.1 (Spring 1982): 18–45.

———. "The Prophet and Her Audience: Gender and Knowledge in the World Turned Upside Down." In *Reviving the English Revolution: Reflections and Elaborations on the Work of Christopher Hill*, eds. Geoff Eley and William Hunt, pp. 139–53. London: Verso, 1988.

———. *Visionary Women: Ecstatic Prophecy in Seventeenth-Century England.* Berkeley: University of California Press, 1992.

Malcovati, H. ed. *Oratorum Romanum Framenta Libera Rei Publicae*, 4th edition. Turin: Paravia, 1976–9, pp. 174–98.

Maley, Willy. "Milton and 'the complication of interests' in Early Modern Ireland." In *Milton and the Imperial Vision*, eds. Balachandra Rajan and Elizabeth Sauer, pp. 155–68. Pittsburgh: Duquesne University Press, 1999.

Marcus, Leah Sinanoglou. *Childhood and Cultural Despair.* Pittsburgh: University of Pittsburgh Press, 1978.

Marvell, Andrew. *The Poems and Letters of Andrew Marvell.* 2 vols. Ed. H. M. Margoliouth. Oxford: Clarendon, 1971.

Matchinski, Megan. *Writing, Gender, and State in Early Modern England: Identity Formation and the Female Subject.* Cambridge: Cambridge University Press, 1998.

Mather, Cotton. *A Father Departing.* Boston: Belnap, 1723.

Mather, Increase. *David Serving His Generation.* Boston: B. Green & J. Allen, 1698. Wing/M1195. *EEBO.*

Maus, Katharine Eisaman. *Inwardness and the Theater in the English Renaissance.* Chicago: University of Chicago Press, 1995.

Mauss, Marcel. *The Gift: Forms and Functions of Exchange in Archaic Societies.* Trans. Ian Cunnison. New York: Norton, 1967.

Mazzeo, Joseph Anthony. *Renaissance and Seventeenth-Century Studies.* New York: Columbia University Press, 1964.

McColley, Diane. "Subsequent or Precedent? Eve as Milton's Defense of Poesie." *Milton Quarterly,* 20 (1986): 132–6.

———. *A Gust for Paradise: Milton's Eden and the Visual Arts.* Urbana: University of Illinois Press, 1993.

McEachern, Claire. *The Poetics of English Nationhood, 1590–1612.* Cambridge: Cambridge University Press, 1996.

McIlwain, Charles Howard, ed. *The Political Works of James I.* Cambridge, Mass: Harvard University Press, 1918.

McKeon, Michael. *The Origins of the English Novel 1600–1740.* Baltimore: Johns Hopkins University Press, 1987.

McLaren, Dorothy. "Fertility, Infant Mortality, and Breast Feeding in the Seventeenth Century." *Medical History,* 22 (1978): 380–8.

McLeod, Bruce. "The 'Lordly Eye': Milton and the Strategic Geography of Empire." In *Milton and the Imperial Vision,* eds. Balachandra Rajan and Elizabeth Sauer, pp. 51–66. Duquesne University Press, 1999.

Menasseh ben Israel, *De Termino Vitae libri Tres.* Amsterdam, 1639. *Of the Term of Life.* London: J. Nutt, 1699.

Mikalachki, Jodi. *The Legacy of Boadicea: Gender and Nation in Early Modern England.* London: Routledge, 1998.

Miller, Jacqueline T. "Mother Tongues: Language and Lactation in Early Modern Literature," *ELR,* 27 (1997): 178.

Miller, Naomi J. "Playing 'the mother's part': Shakespeare's Sonnets and Early Modern Codes of Maternity." In *Shakespeare's Sonnets: Critical Essays,* ed. James Schiffer, pp. 347–68. New York and London: Garland Books, 1999.

Miller, Naomi J. and Naomi Yavneh, eds. *Maternal Measures: Figuring Caregiving in the Early Modern Period.* Burlington, VT: Ashgate, 2000.

Miller, Shannon. *Engendering the Fall: John Milton and Seventeenth-Century Women Writers.* Philadelphia: University of Pennsylvania Press, 2008.

Milton, John. *The Complete Poetry and Essential Prose of John Milton.* Eds. William Kerrigan, John Rumrich, and Stephen M. Fallon. New York: The Modern Library, 2007.

More, Henry. *Enthusiasm triumphatus: or, A discourse on the nature, causes, kinds, and cure of enthusiasme.* London: J. Flesher, 1656. Thomason/E.1580[1]. *EEBO.*

Morrill, John, ed. *Oliver Cromwell and the English Revolution*. London: Longman, 1990.

Morton, Timothy. *The Poetics of Spice: Romantic Consumerism and the Exotic*. Cambridge: Cambridge University Press, 2000.

Mueller, Janel. "Dominion as Domesticity: Milton's Imperial God and the Experience of History." In *Milton and the Imperial Vision*, eds. Balachandra Rajan and Elizabeth Sauer, pp. 25–47. Pittsburgh: Duquesne University Press, 1999.

Nairn, Tom. *The Break-up of Britain: Crisis and Neo-Nationalism*. 2nd ed. London: Verso: 1981.

Norbrook, David. "*Macbeth* and the Politics of Historiography." In *Politics of Discourse: The Literature and History of Seventeenth-Century England*, eds. Kevin Sharpe and Steven N. Zwicker, pp. 78–116. Berkeley: University of California Press, 1987.

——. *Writing the English Republic: Poetry, Rhetoric and Politics, 1627–1660*. Cambridge: Cambridge University Press, 1999.

Nyquist, Mary. "The Genesis of Gendered Subjectivity in the Divorce Tracts and in *Paradise Lost*." In *Re-membering Milton: Essays of the Texts and Traditions*, eds. Mary Nyquist and Margaret Ferguson, pp. 99–127. London: Methuen, 1987.

——. "'Profuse, proud Cleopatra': 'Barbarism' and Female Rule in Early Modern English Republicanism." In *The Representation of Gender in the English Revolution*, ed. Sharon Achinstein. *Women's Studies*, 24 (1994): 85–130.

Oldman, Elizabeth. "Milton Grotius, and the Law of War: A Reading of 'Paradise Regained' and 'Samson Agonistes'." *Studies in Philology*, 104.3 (2007): 340–75.

Olyan, Saul M. "'They Shall Wail the Songs of the Temple': Sanctioned Mourning in Biblical Cultic Settings," paper, University of New Hampshire, 2000.

Orgel, Stephen. "Propero's Wife." In *Rewriting the Renaissance*, eds. Margaret Ferguson, Maureen Quilligan, and Nancy Vickers, pp. 50–64. Chicago: University of Chicago Press, 1986.

——. *Impersonations: The Performance of Gender in Shakespeare's England*. Cambridge: Cambridge University Press, 1996.

Orlin, Lena Cowen. *Private Matters and Public Culture in Post-Reformation England*. Ithaca: Cornell University Press, 1994.

Overbury, Thomas. *New and choise characters*. London: Laurence Lisle, 1615. STC/1732.25. *EEBO*.

Ozment, Steven. *When Fathers Ruled: Family Life in Reformation Europe*. Cambridge, Mass: Harvard University Press, 1983.

Paster, Gail Kerns. *The Body Embarrassed: Drama and the Discipline of Shame in Early Modern England*. Ithaca: Cornell University Press, 1993.

Patterson, Annabel M. *Marvell and the Civic Crown*. Princeton: Princeton University Press, 1978.

Peacock, John. "The Visual Image of Charles I." In *The Royal Image: Representations of Charles I*, ed. Thomas N. Corns, pp. 176–289. Cambridge: Cambridge University Press, 1999.

Peers, Allison, E. ed. and trans. *Saint Teresa, The Complete Works.* 2 vols. London and New York: Sheed and Ward, 1946.

Perry, Ruth. "Colonizing the Breast: Sexuality and Maternity in Eighteenth-Century England." *Eighteenth-Century Life,* 16 (1992): 194.

Peters, Julie Stone. "'A Bridge Over Chaos': *De Jure Belli, Paradise Lost,* Sovereignty, Globalism, and the Modern Law of Nations." *Comparative Literature,* 57.4 (Fall 2005): 273–93.

Phayer [Phaer], Thomas. *The kegiment[sic] of life whereunto is added a treatyse of the pestilence, with The book of children newly corrected and enlarged by T. Phayer.* London: Edward Whitchurche, 1546. STC 2nd ed/11969. *EEBO.*

Philo. *Loeb Classics Philo.* London: Heinemann, 1932.

Piciotto, Joannna. "The Public Person and the Play of Fact." *Representations.* 105.1 (February 2009): 85–132.

Playfere, Thomas. *The Meane in Mourning.* STC, 20017. London: James Robert, 1597. STC/20017. *EEBO.*

Pocock, J.G.A. "British History: A Plea for a New Subject." *Journal of Modern History,* 7 (1975): 601–21.

Popkin, Richard H. "Jewish Messianism and Christian Millenarianisn." In *Culture and Politics from Puritanism to the Enlightenment,* ed. Perez Zagorin, pp. 67–90. Berkeley and Los Angeles: University of California Press, 1980.

Powell, Chilton Latham. *English Domestic Relations 1487–1653: A Study of Matrimony and Family Life in Theory and Practice As Revealed by the Literature, Law, and History of the Period.* New York: Columbia University Press, 1917.

Pricke, Robert. *The doctrine of superiority.* London: [T. Creede], 1609. STC (2nd ed.)/20337. *EEBO.*

Purchas, Samuel. *Purchas His Pilgrims, or Hakluytus Posthumus.* 20 vols. Glasgow: James Maclehose and Sons, 1940.

Purkiss, Diane. "Material Girls: The Seventeenth-Century Woman Debate." In *Women, Texts and Histories 1575–1760,* eds. Clare Brant and Diane Purkiss pp. 69–100. London: Routledge, 1992.

——. "Producing the Voice, Consuming the Body: Women Prophets of the Seventeenth Century." In *Women, Writing, History, 1640–1740,* eds. Isabel Crundy and Susan Wiseman. pp. 139–58. Athens, Georgia: University of Georgia Press, 1992.

——. "*Macbeth* and the All-singing, All-dancing Plays of the Jacobean Witchvogue," in *Shakespeare, Feminism, and Gender,* ed. Kate Chedgzoy. pp. 216–34 New York: Palgrave, 2001.

Quilligan, Maureen. *Milton's Spenser: The Politics of Reading.* Ithaca: Cornell University Press, 1983.

Quint, David. *Epic and Empire: Politics and Generic Form from Virgil to Milton.* Princeton: Princeton University Press, 1993.

Ragussis, Micahel. *Figures of Conversion: "The Jewish Question" and English National Identity.* Durham: Duke University Press, 1996.

Rajan, Balachandra. "Banyan Trees and Fig Leaves: Some Thoughts on Milton's India." In *Of Poetry and Politics: New Essays on Milton and His World,* ed. P.G. Stanwood. Binghamton, NY: Medieval and Renaissance Texts and Studies, 1995.

Raymond, Joad. "Popular Representations of Charles I." In *The Royal Image: Representations of Charles I*, ed. Thomas N. Corns, pp. 47–73. Cambridge: Cambridge University Press, 1999.

——. "Complications of Interest: Milton, Scotland, Ireland, and National Identity in 1649." *Review of English Studies* (2004): 315–45.

Reiss, Timothy. "Montaigne and the Subject of Polity." In *Literary Theory/Renaissance Texts*, eds. Patricia Parker and David Quint, pp. 115–49. Baltimore: The Johns Hopkins University Press, 1986.

Richards, Judith M. "The English Accession of James VI: 'National' Identity, Gender and the Personal Monarchy of England." *The English Historical Review*, 117.472 (2002): 513–35.

Robertson, Geoffrey. *The Tyrannicide Brief*. London: Chatto & Windham, 2005.

Rogers, Daniel. *Matrimoniall Honour: or The mutuall Crowne and comfort of godly, loyall, and chaste Marriage*. London: Thomas Harper, 1642.

Rogers, John. *The Matter of Revolution: Science, Poetry, and Politics in the Age of Milton*. Ithaca: Cornell University Press, 1996.

Romack, Katherine. "Monstrous Births and the Body Politic: Women's Political Writings and the Strange and Wonderful Travails of Mistris Parliament and Mris. Rump." In *Debating Gender in Early Modern England, 1500–1700*, eds. Christina Malcolmson and Mihoko Suzuki, pp. 209–30. New York and London: Palgrave, 2002.

Rose, Mary Beth. *Gender and Heroism in Early Modern Literature*. Chicago: University of Chicago Press, 2006.

Rosenblatt, Jason. *Torah and Law in Paradise Lost*. Princeton: Princeton University Press, 1994.

——. *Renaissance England's Chief Rabbi: John Selden*. Oxford: Oxford University Press, 2006.

Rosenfild, Kirstie Gulick. "Nursing Nothing: Witchcraft and Female Sexuality in *The Winter's Tale*." *Mosaic*, 35 (2002): 95–112.

Russell, Conrad, ed. *The Origins of the English Civil War*. New York: Barnes & Noble, 1973.

——. *The Causes of the English Civil War*. Oxford: Clarendon Press, 1990.

Sand, Shlomo. *The Invention of the Jewish People*. Trans. Yael Lotan. London: Verso, 2009.

Sauer, Elizabeth. "Maternity, Prophecy, and the Cultivation of the Public Sphere in the Seventeenth Century." *Renaissance Culture*, 24 (1998): 118–48.

——. "Religious Toleration and Imperial Intolerance." *Milton and the Imperial Vision*, pp. 214–32, eds. Balachandra Rajan and Elizabeth Sauer. Pittsburgh: Duquesne University Press, 1999.

——. "The Experience of Defeat: Milton and Some Female Contemporaries." In *Milton and Gender*, ed. Catherine Gimelli Martin, pp. 135–52. Cambridge: Cambridge University Press, 2004.

——. "Milton's *Of True Religion*, Protestant Nationhood, and the Negotiation of Liberty." *Milton Quarterly*, 40.1 (2006): 1–19.

Schnell, Lisa J. "Muzzling the Competition: Rachel Speght and the Economics of Print." In *Debating Gender in Early Modern England, 1500–1700*, eds. Christina Malcolmson and Mihoko Suzuki, pp. 57–78. New York: Palgrave, 2002.

Schnucker, R.V. "The English Puritans, Delivery and Breast-Feeding," *History of Childhood Quarterly*, 1 (1974): 637–58.

Schwartz, Louis. *Milton and Maternal Mortality.* Cambridge: Cambridge University Press, 2009.

Schwarz, Katherine. "Missing the Breast: Desire, Disease, and the Singular Effect of Amazons." In *Body in Parts: Fantasies of Corporeality in Early Modern Europe*, eds. David Hillman and Carla Mazzio, pp. 147–69. New York: Routledge, 1997.

Schwyzer, Philip. "Phaer, Thomas (1510?–1560)." *Oxford Dictionary of National Biography*; online ed. October 2009: http://www.Oxforddnb.com/view/article/22085.

Selden, John. *Ioannis Seldeni Uxor Ebraica, seu De Nuptiis & Divortiis ex Iure Civili, Id Est, Divino & Talmudico, Veterum Ebraeorum.* London: Typis Richardi Bishopii, 1646. Wing/S2443. *EEBO.*

Shapiro, James. *Shakespeare and the Jews.* New York: Columbia University Press, 1996.

Sharp, Jane. *The Midwives Book.* London, 1671. Wing/S2969B. *EEBO.*

Sharpe, Kevin. *The Personal Rule of Charles I.* New Haven: Yale University Press, 1992.

——. "The Royal Image: An Afterward." In *The Royal Image: Representations of Charles I*, ed. Thomas N. Corns, pp. 288–30. Cambridge: Cambridge University Press, 1999.

Sharpe, Kevin and Peter Lake, eds. *Culture and Politics in Early Stuart England.* Palo Alto: Stanford University Press, 1994.

Shawcross, John. *John Milton: The Self and the World.* Lexington: University Press of Kentucky, 1993.

Shell, Marc. "Marranos (Pigs); or From Coexistence to Toleration." *Critical Inquiry*, 17 (1991): 306–36.

Shoemaker, Robert B. *Gender in English Society, 1650–1850: The Emergence of Separate Spheres.* London: Addison Wesley Longman Limited, 1998.

Shoulson, Jeffrey. *Milton and the Rabbis.* New York: Columbia University Press, 2001.

Shuger, Deborah. *Habits of Thought in the English Renaissance: Religion, Politics, and the Dominant Culture.* Berkeley: University of California Press, 1990.

Silver, Victoria. *Imperfect Sense: The Predicament of Milton's Irony.* Princeton: Princeton University Press, 2001.

Skinner, Cyriack. "The Life of Mr. John Milton." In *The Riverside Milton*, ed. Roy Flannagan. Boston: Houghton Mifflin, 1998.

Smith, Nigel. *Perfection Proclaimed: Language and Literature in English Radical Religion 1640–1660.* Oxford: Clarendon Press, 1989.

——. "Popular Republicanism in the 1650s: John Streater's 'Heroick Mechanicks'." In *Milton and Republicanism*, eds. David Armitage, Arman Himy, and

Quentin Skinner, pp. 137–55. Cambridge: Cambridge University Press, 1995.

——. *Literature and Revolution in England 1640–1660*. New Haven: Yale University Press, 1997.

Somerville, M.A., David. *St. Paul's Conception of Christ or The Doctrine of the Second Adam*. Edinburgh: T&T Clark, 1897.

Sowerman, Esther. *Ester hath hang'd Haman or An Answer to a lewd Pamphlet, entitled The Arraignment of Women*. STC/22974. London, 1617.

Spenser, Edmund. *A View of the Present State of Ireland*. Ed. W. L. Renwick. Oxford: Clarendon, 1970.

Spurr, John. "Earle, John (1598–1601–1665)." *ODNB*, online edn, January 2008: http://www.oxforddnb.com/view/article/8400.

Steadman, John M. *The Hill and the Labyrinth: Discourse and Certainty in Milton and His Near-Contemporaries*. Berkeley: University of California Press, 1984.

Stevens, Paul. "'Leviticus Thinking' and the Rhetoric of Early Modern Colonialism." *Criticism*, 35 (1993): 441–61.

——. "Spenser and Milton on Ireland: Civility, Exclusion, and the Politics of Wisdom." *Ariel*, 26 (1995): 151–67.

——. "Milton's Janus-faced Nationalism: Soliloquy, Subject, and the Modern Nation State." *JEGP*, 100.2 (2001): 264–8.

——. "How Milton's Nationalism Works: Globalization and the Possibilities of Positive Nationalism." In *Early Modern Nationalism and Milton's Nation*, eds. David Loewenstein and Paul Stevens, pp. 273–304. Toronto: University of Toronto Press, 2008.

Stevenson, David. *Revolution and Counter-Revolution in Scotland, 1644–51*, 1977, rpt. Edinburgh: John Donald-Birlinn, 2003.

Stone, Lawrence. *The Family, Sex and Marriage in England, 1500–1800*. New York: Harper & Row, 1977.

Strachniewski, John. *The Persecutory Imagination: English Puritanism and the Literature of Religious Despair*. Oxford: Clarendon Press, 1991.

Streater, John. *Observations, Historical, Political and Philosophical, upon Aristotle's first Book of Political Government*. 11 nos. London: John Streater, 4 April–4 July 1654.

Strong, Roy. *Van Dyck: Charles on Horseback*. London: Allen Lane, 1972.

Suzuki, Mihoko. *Subordinate Subjects: Gender, the Political Nation, and Literary Form in England, 1588–1688*. Burlington, VT: Ashgate, 2002.

Tayler, Edward W. *Milton's Poetry: Its Development in Time*. Pittsburgh: Duquesne University Press, 1979.

Thickstun, Margaret. "Milton Among Puritan Women: Affiliative Spirituality and the Conclusion of 'Paradise Lost'." *Religion and Literature*, 36.2 (Summer 2004): 1–23.

Thomas, Keith. "Women and the Civil War Sects." *Past and Present*, 13 (1958): 42–62.

Trible, Phillis. *God and the Rhetoric of Sexuality*. Minneapolis: Fortress, 1978.

Trubowitz, Rachel. "Female Preachers and Male Wives: Gender and Authority in Civil War England." In *Pamphlet Wars: Prose in the English Revolution*, ed. James Holstun, pp. 112–33. London: Frank Cass, 1992.

——. "'The people of Asia and with them the Jews': Israel, Asia, and England in Milton's Writings." In *Milton and the Jews*, ed. Douglas Brooks, pp. 151–74. Cambridge: Cambridge University Press, 2008.

——. "Sublime Pauline: Denying Death in *Paradise Lost*." In *Imagining Death in Spenser and Milton*, eds. Elizabeth Jane Bellamy, Patrick Cheney, and Michael Schoenfeldt, pp. 131–50. New York: Palgrave, 2003.

Trumback, Randolph. *The Rise of the Egalitarian Family*. New York: Academic Press, 1978.

Turner, James Grantham. *One Flesh Paradisal Marriage and Sexual Relations in the Age of Milton*. Oxford: Clarendon Press, 1987.

Vane, Henry Sir. *A Needful Corrective or Balance in Popular Government*. London, 1660. Wing/V72. *EEBO*.

Veldman, Ilja M. "Who is the Strongest? The Riddle of Esdras in Netherlandish Art." *Simiolus: Netherlands Quarterly for the History of Art*, 17.4 (1987): 223–9.

Vicary, Thomas. *The Anatomie of the Bodie of Man*, 1548. Eds. Frederick J. Furnivall and Percy Furnivall. Reissued by the Surgeons of St. Bartholomew's in 1577. London: Early English Texts Society, 1888.

Viswanathan, Gauri. "Milton and Education." In *Milton and the Imperial Vision*, eds. Balachandra Rajan and Elizabeth Sauer, pp. 273–93. Pittsburgh: Duquesne University Press, 1999.

Walker, William. "Antiformalism, Antimonarchism, and Republicanism in Milton's 'Regicide Tracts'." *Modern Philology*, 108.4 (May 2011): 507–37.

Wall, Wendy. *Staging Domesticity: Household Work and English Identity in Early Modern Drama*. Cambridge: Cambridge University Press, 2002.

Walsh, James, ed. and trans. *Julian of Norwich, Revelations of Divine Law*. New York: Harper, 1961.

Walzer, Michael. *Exodus and Revolution*. New York: Basic Books, 1985.

Warburton, Rachel. "Mary Cary (Rande), *Little Horns Doom and Downfall and a New and More Exact Mappe of the New Jerusalems Glory* (1651)." In *Reading Early Modern Women: an Anthology of Texts in Manuscript and Print 1550–1700*, eds. Helen Ostovich and Elizabeth Sauer, pp. 138–9. London: Routledge, 2004.

Warner, Marina. *Alone of All Her Sex: The Myth and Cult of the Virgin Mary*. New York: Vintage, 2000.

Watson, Robert. *The Rest is Silence: Death as Annihilation in the English Renaissance*. Berkeley: University of California Press, 1994.

Wayne, Valerie. "Advice for Women from Mothers and Patriarchs." In *Women and Literature in Britain, 1500–1700*, ed. Helen Wilcox, pp. 56–79. Cambridge: Cambridge University Press, 1996.

Webber, Joan. *The Eloquent "I": Style and Self in Seventeenth-Century Prose*. Madison: University of Wisconsin Press, 1968.

Whatley, Christopher A. *The Scots and the Union*. Edinburgh: Edinburgh University Press, 2006.

Wheeler, Elizabeth Skerpan. "*Eikon Basilike* and the Rhetoric of Self-Representation." In *The Royal Image: Representations of Charles I*, ed. Thomas Corns, pp. 122–40. Cambridge: Cambridge University Press, 1999.

Wilcher, Robert. *The Writing of Royalism, 1628–1660*. Cambridge: Cambridge University Press, 2001.

Willard, Samuel. *A Compleat Body of Divinity*. Boston, 1726.

Williams, Raymond. *In The Year 2000*. New York: Pantheon, 1983.

Willis, Deborah. *Malevolent Nurture: Witch-Hunting and Maternal Power in Early Modern England*. Ithaca: Cornell University Press, 1995.

Willoughby, Percivall. *Observations in Midwifery*. First printed and edited by Henry Blenkinsop (1893). Reprinted with Introduction by John L. Thronton. Yorkshire: S.R. Publishers, 1972.

Wilson, Richard. "Observations on English Bodies: Licensing Maternity in Shakespeare's Late Plays." In *Enclosure Acts: Sexuality, Property, and Culture in Early Modern England*, eds. Richard Burt and John Michael Archer, pp. 121–50. Ithaca: Cornell University Press, 1994.

Winstanley, Gerrard. *The Works of Gerrard Winstanley*. Ed. George Sabine. Ithaca: Cornell University Press, 1941.

Wiseman, Susan. "Unsilent Instruments and the Devil's Cushion: Authority in Seventeenth-Century Women's Prophetic Discourse." In *New Feminist Discourse: Critical Essays on Theories and Texts*, ed. Isobel Armstrong, pp. 176–96. London: Routledge, 1992.

Wittreich, Joseph. *Interpreting Samson Agonistes*. Princeton: Princeton University Press, 1986.

——. *Shifting Contexts: Reinterpreting Samson Agonistes*. Pittsburgh: Duquesne University Press, 2002.

Wolfe, Don, M., gen.ed. *The Complete Prose Works of John Milton*. 8 vols. New Haven: Yale University Press, 1953–82.

Worden, Blair. "Andrew Marvell, Oliver Cromwell, and the *Horatian Ode*." In *Politics of Discourse*, eds. Kevin Sharpe and Steven N. Zwicker, pp. 147–80. Berkeley: University of California Press, 1987.

Yalom, Marilyn. *A History of the Breast*. New York: Knopf, 1997.

Young, Robert J.C. "The Overwritten Underwritten: Nationalism and its Doubles in Postcolonial Theory." In *The Silent Word: Textual Meaning and the Underwritten*, eds. Robert Young, Ban Kah Choon, and Robbie B.H. Gobb. Introduction. Singapore: University of Singapore Press, 1998.

Zwicker, Steven. "Models of Governance in Marvell's 'The First Anniversary'." *Criticism*, 16 (1974): 1–12.

Index

Printed and bound by CPI Group (UK) Ltd, Croydon, CR0 4YY